Fred Heman Hubbard

Opium Habit and Alcoholism

A Treatise on the Habits of Opium and Its Compounds

Fred Heman Hubbard

Opium Habit and Alcoholism
A Treatise on the Habits of Opium and Its Compounds

ISBN/EAN: 9783337138585

Printed in Europe, USA, Canada, Australia, Japan

Cover: Foto ©berggeist007 / pixelio.de

More available books at **www.hansebooks.com**

THE OPIUM HABIT

AND

ALCOHOLISM.

A TREATISE ON

THE HABITS OF OPIUM AND ITS COMPOUNDS;
ALCOHOL; CHLORAL-HYDRATE; CHLORO-
FORM; BROMIDE POTASSIUM;
AND CANNABIS INDICA:

INCLUDING THEIR

THERAPEUTICAL INDICATIONS:

WITH SUGGESTIONS FOR TREATING VARIOUS

PAINFUL COMPLICATIONS.

BY

DR. FRED. HEMAN HUBBARD.

———

A. S. BARNES & CO.,
111 AND 113 WILLIAM STREET,
NEW YORK.

TO

LOUIS P. TWYEFFORT,

THIS VOLUME IS RESPECTFULLY AND AFFECTIONATELY
DEDICATED BY HIS GRATEFUL FRIEND.

THE AUTHOR.

THE OPIUM HABIT AND ALCOHOLISM.

INTRODUCTORY.

IN writing a memoir on the diseased state of the system engendered by the habitual use of powerful drugs and stimulating liquors, and indicating a rational treatment for the same, the author has kept one object steadily in view; he has sought to make his work useful, and to place in the hands of the profession a carefully arranged analysis of the peculiar physical condition induced by such indulgence, a condition which makes necessary each day a certain measure of stimulative, to sustain the system in its abnormal state.

We shall treat each habit separately, describing the abnormal conditions peculiar to each, and defining their essential characteristics. We shall minutely describe the symptoms that appear, and the changes that take place under treatment; so that the practitioner may familiarize himself with them, while following this record of the results of our twelve years' observation. While treating this class of patients in asylums and private practice, the author has exercised scrupulous care in noting each peculiarity, and the treatment which proved successful in relieving each. The causes leading to the habitual use of the narcotics are many, and the more important of them will receive due attention.

We shall not refer to the moral aspect of our subject; that would not be consistent with the scope of this work. Neither shall we speculate on the effects which the opium and alcohol habits are likely to have on the descendants of their victims, except as the matter presents itself from the physician's point of view in the cases of mothers who are addicted to opium during the child-bearing period. For such cases we suggest the treatment necessary to save both mother and child.

The pernicious habit of taking opium or its compounds has rapidly increased since the introduction of the hypodermic syringe. It now counts its victims by thousands. The medical profession has been strangely apathetic with respect to the increase of this vice. The result has been to leave in the hands of unprincipled quacks the treatment of a disease, which involves marked pathological conditions, and requires for its successful treatment an acquaintance with the details of physiological anatomy, and a perfect familiarity with certain striking characteristics of the disease which appear in individual cases.

The unfortunate patients, in their eagerness for relief. have become willing victims, enticed by the deceptive reports and advertisements of the many asylums and sanitariums which flood the country.

The habitual use of opium is a disease, and a formidable one; and, without regard to the labor involved, the author has endeavored to present in a concise form the results of his experience and observation from day to day; giving exact reports of observations made during the treatment of cases. He has selected for this purpose typical cases, having interesting histories and complications, the patients being of all ages and conditions, from the babe in arms to the octogenarian. Great care has been taken in describing the various peculiarities developed in different patients by the use of the drug.

The habit, while holding its victims in a degrading slavery, is amenable to treatment. The mental capabilities return, when the pathological states created by the opium have been removed. The influence exerted by opium over diathetic tendencies, and its power to suppress and control diseases of a certain character, is of paramount importance to the physician. We have, therefore, given this branch of the subject that careful consideration which it demands. Prostrating diseases, involving the nervous system, predispose the sufferer to indulge in powerful stimulants. Such diseases interfere also with a successful result, when the effects of habit are under treatment, whether the disease is present as a concomitant of the habit, or as the exciting cause of its formation. It becomes absolutely necessary, therefore, to cure or to control the activity of any accompanying complications of this nature. To this end we have included the treatment of painful diseases as matter necessary to be considered in this treatise. Taking first the habitual use of opium and its compounds, we treat of the peculiarities of the habit, following with the treatment of individual cases. Catarrh, rheumatism, pulmonary tuberculosis, cancer of the uterus, with inflammation and ulceration of the cervix, have been suffered by patients under the author's treatment for the opium habit. We refer to the methods used for their relief, and the results attained. The influence exerted by opium over the uterus and its power to suppress the catamenics and the lactiferous secretions, present an interesting anomaly, and will receive attention accordingly. Dyspepsia, neuralgia and sciatica, as modified by the abuse of stimulants, are discussed, with their treatment, independent of dementing drugs. Other topics are: Evil results of administering paregoric, laudanum, soothing syrup, or black drops to children, the means by which their use may be avoided, with a cursory suggestion, relating to the treatment of childrens' diseases during the summer months; opium stimulation, with differential points in diagnosis, between the effects of opium and of liquor, as compared with the delirium of disease, including full descriptions of the symptoms peculiar to each condition, and the hypodermic injection of morphia its dangers, complications and tendencies, and the means of avoiding its use to a certain extent, in practice. Other themes will be considered as follows

Bromide potassium· Its dementing effects, when taken habitually, the power exerted by the drug over epilepsy Its antaphrodisiac properties, how it may be administered, and the treatment of the bromide habit. Chloral. Its treacherous and subtle tendencies, and the fatal results following its administration, post-mortem changes.

Chloroform· How its dangerous effects may be warded off, Chloroform intoxication, and its results; accidents incident to the inhalation of chloroform; how to create a disgust for its use. A preparatory treatment for pregnant women, to modify or entirely allay the pains of parturition; the treatment requisite to prepare mother and fœtus, by a thorough course of hygienic measures (with dietary suggestions), obviating the necessity for using anœsthetics during labor. This treatment also removes one of the principal incentives influencing women to commit the horrible crime of abortion.

We shall next consider the habit of indulging in intoxicating drinks, and the disease dipsomania, or an irresistible longing for liquor. The pathology and treatment of this disease, under the many phases in which it is presented to the physician in general practice, will be taken up in detail, with respect to morning and evening dram drinking its effects in times past, as compared with the present; the every day drinker who does not become intoxicated, yet indulges constantly; the periodical drunkard, and the methods to pursue in stopping his debauch; also delirium tremens : its cause, pathology and treatment.

The subject of favorable surroundings for patients, while undergoing treatment, we shall next consider, and inquire, why inebriate asylums and sanitariums have made so signal a failure in their efforts to reform the fallen.

The author presents for careful consideration a subject of grave importance, as it involves the life and happiness of many persons who are now approaching the threshold of manhood, and are soon to be exposed to the temptations incident to that estate. Whether it is our duty to presuppose a remote contingency and subject young men to treatment for that which has not yet developed itself, is a question which must be conscientiously decided by the physician, after a searching review of all the facts. If by treatment we cause positive proximate evils, that are apt to endanger our patient's present or future health, that fact constitutes a strong argument against such a course. But no such results follow; and the permanent benefits and immunities to be gained, justify us in exciting the momentary pain necessary to create a distaste for alcoholic drinks, and forever to destroy the ability to consume them, thereby securing to the patient a life of sobriety. When we call to mind the misery and tears of agonized wives and heart-broken parents, we feel compelled to believe that it would be better for nine hundred and ninety-nine young men to be subjected to the treatment which destroys the appetite for liquor, though in their case unnecessary, than to omit treating one, to whom treatment would ensure a life of sobriety.

Statistical history regarding our subject would fill many chapters with interesting matter; but it would not serve the end or purpose for which this memoir is written. As a suggestive fact, showing the rapid increase of the opium habit, we give the official returns of the amount of the drug imported during 1867 and 1877. During 1867 130,105 lbs. and during 1877 230,102 lbs., with 47,028 lbs. for smoking purposes, were imported.

In treating the results of excess, the importance of paying strict attention to minute details should be recognized, as a successful result depends upon the systematic manner in which the order of treatment is carried out. To the physician it should be said: Take into careful consideration the age, surroundings, temperament and idiosyncracies of your patient, and the incentive that first induced a tendency to 'fixed habit. Do not look upon the victims of habit as poor, miserable outcasts, unfit for the society of their more fortunate friends; for they will intuitively read your thoughts, and with one stroke you lose an essential prerequisite to perfect success—their confidence. Remember also, to their credit, that they are not all free and willing devotees at the devil's shrine of opium or liquor. In many cases, acute pain and pardonable ignorance regarding the final effects of intemperance, have placed them inadvertently in its power. Ninety-nine out of every one hundred opium eaters, so called, who have been cultivating the habit during more than three years, would, if it were in their power, throw off the habit willingly by sacrificing a limb; and they will assist you in every way possible, to the full extent of their weakened will-power.

REMARK.

I take this opportunity to inform my readers, that a growing general practice has made it necessary for me to prosecute my labor during the limited time at my disposal between calls. Any one similarly situated, will appreciate the difficulty I have experienced in preparing a connected history of my methods.

Very respectfully,

52 *Monroe Street,*   THE AUTHOR.

Brooklyn, N. Y.

OPIUM HABIT AND ALCOHOLISM.

CHAPTER I.

PECULIARITIES OF THE OPIUM HABIT.

THE power exerted over the human system by opium habitually used, is difficult to describe. No words can adequately express the horrible sufferings endured by the confirmed opium eater. The habit holds the system in shackles as of steel. The subtle influence of the drug is felt in every fibre of the nervous system, creating decided changes in the endangium and neurilemma. The patient finds it impossible to abstain from the poison beyond a fixed time. If an attempt is made to decrease or discontinue its use, complications result which are terrible to endure ; a spasm of the nerves' periphery follows, with complete relaxation of the system. The brain is progressively injured and rendered anæmic for the want of nutrition ; yet the ability to suffer is not diminished.

The dipsomaniac's brain, by continual congestion, is dulled, and his sensibilities are deadened by the liquor, a result which mitigates his sufferings to

a certain extent. The opium eater has no such relief. The opium habit is a disease of the nervous
system, and is not subject to control by the will. In
examining the phenomena of the disease, we cannot
reconcile them with the idea that it is a mere habit, indulged with a desire to stimulate, and in satisfaction of
a depraved appetite. This supposition will not stand the
test of careful investigation, as the facts to be presented
demonstrate. Women of culture and natural refinement
pawn their jewels or sell the feathers from their beds, to
secure the requisite amount of opium. The drunkard,
when under the influence of liquor, will go to great extremes and commit desperate acts to obtain rum ; but
the victim of opium is not actively stimulated, when he
barters his honor for the drug. He knows, however, that
the time is near at hand when the drug must be taken to
sustain the nervous system and ward off prostration
which would entail with unerring certainty the agony of
a thousand deaths. A mere appetite would not drive
him to such extremities.

When the habit is forming, and before any pathological changes have taken place from the continued use of
opium, the effects are exceedingly pleasant and fascinat-
ing, and the opium eater may be tempted to repeat the
dose for the purpose of renewing those effects. No
imperative demand for more is made by the system, but
the victim is enjoying the full exhilarating action of the
drug. When the habit is formed, its delights are lost.
The time has now passed in which to quit the practice,

and the opium eater henceforth derives only misery from a habit which he cannot control. The drug exercises almost a demoniac influence in misleading its victims, inspiring them with conflicting fears, and leading them on to unnecessary excesses by its deceptive effects. While the habit is forming, and before indulgence becomes a necessity to the system, the effect of opium is peculiarly soothing and tranquillizing, the stimulation reaching a certain point and remaining stationary for many hours.

The stimulation of the intellectual powers is extraordinary, producing a harmonious blending of the conceptions.

The effects of liquor or wine, as compared with those of opium, are coarse and brutalizing. Alcoholic intoxication does not induce the state of waking dreams, inspired by opium, which carries the mind gently through fairy-land, exaggerating all that is sublime in the opium eater's nature, allaying pain and drowning trouble in a sense of the beautiful. In these earlier stages of opium eating the imagination is stimulated to wonderful activity. Sights rare and pleasing pass kaleidoscopically through the mind, each bringing with it some new delight of the senses. As sleep approaches, a state of perfect rest ensues; and for hours preceding reaction, a semi-unconscious state is enjoyed, which surpasses all other sensations in its voluptuous delights. By the first decided effects of the drug, the nerves' periphery is played upon; all that is beautiful presents itself to the imagination.

This is but for a time, however. With the coming of reaction, the victim is racked with spasms, his nerves enduring the extremity of agony. A transfer to the rack or the stake would be a welcome change. During the first entrancing effects of the indulgence, the sense of time is lost in the delights of the imagination, and the pleasures of a month are experienced in an hour. In like manner, when the period of suffering comes, this effect is present, and the sufferer imagines that hours have been passed in this exquisite hell of torment, when in reality only a few minutes have elapsed.

The drug misleads its victims, regarding their own ability to maintain mastery over it. Each individual is confident that in his case there is no danger of forming a habit. There is no shadow cast before, as a warning of the coming misery. Pains are soothed; life seems to possess new charms; the victims are reluctant to believe that the action of the poison will ever be otherwise than pleasant, or to doubt their own ability to stop its use at will.

But alas! very few ever will that it shall be stopped; they continue the indulgence with one more dose, repeated until the fetters gall them. It is then too late; the die is cast, and they recognize the hideous truth that a fixed habit has been formed.

The physiological action of opium is marked and general, presenting, however, but few local manifestations that indicate the profound effect the drug is having on the system. The secretions are diminished, and the activity of the absorbents lessened; constipation is produced through the want of a lubricating material, normally supplied by a secretion from the alimentary tract; the canal is also wanting in tone sufficient to excite peristaltic action.   The secretion of urine is diminished, although frequent libations of water are taken during the first stages of the poisoning, owing to the action of opium upon the blood.   The opium superinduces a thickened state of the fibrin, and brings about an accumulation of effete matter, which renders functional action through the medulla-oblongata slow and laborious, and creates a demand for water by drying the mucous coats of the mouth until the constituent parts of the blood are made normal again by the free consumption of water.   The bladder is unable to discharge its contents, owing to a contracted state of the urethra in its visceral passage, and not because of a paralysis of its muscular walls, as many writers persist in declaring. These muscles are brought into active play, the effort being complete and forcible, as is shown by the fact that oblique inguinal hernia is caused by the effort.   With all the force brought to bear upon the bladder, the stream is small.   The mucous coats of the rectum are dry and hot, making defecation difficult and painful, and often producing hemorrhoids.   The ability or desire to accomplish the sexual act has been completely destroyed in ninety-eight per cent. of all cases observed by us.   After the habit has been formed, and the system has adapted itself to its effects, the physiological conditions, referred to above, are greatly modified; but the functions do not assume a normal state during the continuance of the indulgence.

The victim of opium is bound to a drug from which he derives no benefits, but which slowly deprives him of health and happiness, finally to end in idiocy or premature death. Whatever the victim's condition or surroundings may be, the opium must be taken at certain times with inexorable regularity. The liquor or tobacco user can, for a time, go without the use of these agents, and no regular hours are necessary. During sickness, and more especially during the eruptive fevers, he does not desire tobacco or liquor. The opium eater has no such reprieves; his dose must be taken, and, in painful complications affecting the stomach, a large increase is demanded to sustain the system.

If, in forming the habit, two doses are taken each day, the victim is obliged to maintain that number. It is the unceasing, everlasting slavery of regularity that humiliates opium eaters by a sense of their own weakness. We will note a case, showing the force of the habit and the fatal results that followed its sudden discontinuance. A married lady, twenty-two years of age, residing in a small village in Kentucky, formed the habit in consequence of a painful uterine disease. Her husband, not appreciating the necessity of the drug for his wife, commanded its immediate discontinuance. She attempted to reason with him, but to no purpose; he would accept no explanation, but reiterated his command. The wife knew how impossible it would be to obey, and succeeded in deceiving her husband for some time. On discovering her supposed determination to indulge in the use of opium for its stimulating effects, and her duplicity in deceiving him, his anger was excessive. He notified the druggists with threats, not to sell opium to his wife, or to any one else for her ; and did not consider the matter of sufficient importance to make it necessary that he should inform her of what he had done. She usually obtained the drug as she needed it, keeping no supply on hand. During the day, the young wife, heavy with child, seemed to have a presentiment of the terrible fate in store for her. She said to a companion : "I have an awful fear that my husband will do something desperate ; if he stops my morphine I shall surely die." She discovered what course he had taken to make her stop by the druggist's refusal to sell her opium. It was now time for her evening dose, and relaxation, induced by undue excitement and worry, began to show itself. Knowing her husband's character, she was confident that an attempt to get the drug in town would be useless. She therefore ordered a black boy to mount a fleet horse and go to a village five miles distant to obtain it. During his absence she was seen by the neighbors to stand statute like, holding to the gate, looking intently in the direction the boy had taken. On his return she rushed wildly out, and eagerly listening to what he said, she turned, uttering a loud shriek, and fell down insensible. The neighbors carried her into the house and hurriedly sent for a physician, who found that two lives had been sacrificed to the opium, for the woman was dead. Her husband had notified the distant druggist, and the boy could not obtain the opium. The woman's system being relax for the want of its usual support, and the circulation rapid, so sudden a revulsion of feeling, while the patient was in a delicate state, and in no condition to receive and sustain a shock, produced without doubt valvular insufficiency, or an extravasation of blood into the brain substance.

While understanding the nature and power of the drug, from experience, the opium eater is ignorant of its remote or obscure effects. Its action inspires in its victims a profound fear of fatal results, if they do not continue to take a certain amount. They give the drug credit for greater powers than it possesses. They dare not reduce the quantity taken, through fear of great suffering or death, when in reality no such results would follow, except in rare instances, when the patient is laboring under great fear, or has an organic disease of the heart; and exceptions are useless as precedents. Victims rarely attempt a self-cure, by reducing their dose; for on the first approach of relaxing pain, they resort at once to the poison without stopping to do battle against their mortal enemy. When the system has adapted itself to the drug, and requires a certain amount of it to support and maintain the abnormal condition created by its use, the natural tendency is to increase the dose. Could opium eaters control the desire for this increase, no great or rapid changes would take place; but a large majority surrender to this supposed demand, and soon come to consume prodigious quantities of the drug. Disagreeable sinking sensations, with anorexia and lassitude, and hot and cold flashes, are prominent symptoms that influence them to increase the dose. When the increase is made, it is like ascending a ladder and casting away the rounds; the victims seldom return to their former basis. The larger the dose consumed, the more the system seems to demand, as the drug multiplies complications. The patients suffer intensely, under its debilitating effect; a sleepy habitude ensues, accompanied by a nauseating fulness, difficult to describe. A larger dose is taken in a desperate effort to allay disagreeable sensations, and this process is continued until the system can tolerate no more without exciting alarming symptoms of narcotism, yet the victims suffer apparently for more, when in reality the suffering is the legitimate result of excessive doses. If they would reduce the amount taken to one-half, relief would follow at once, and their sufferings would be comparatively light to bear.

The ability of the system to tolerate large doses, and their effect, differs greatly with different persons. We have cured patients, who, for years, had consumed one drachm of morphia per day; others, taking fifteen grains during the same length of time, would suffer as acutely as the first. One of our patients, a lawyer by profession, was obliged to have an amanuensis to take important

notes in court, as he could not possibly keep awake for
any length of time, consuming daily, as he did, fifteen
grains of morphia hypodermically administered.

The finer sensibilities are generally suppressed. Patients act and appear odd, are prone to exaggerations
and unreasonable statements. They generally have some
utopian scheme in hand, engrossing their attention, of
the success of which they are sanguine. They are weak
and anæmic, with a jaundiced, thickened complexion
and sunken eyes. In seventy-nine per cent. of the cases
coming under our observation, progressive loss of flesh
followed the habitual use of opium, owing to an inability
upon the part of the absorbents to take up nutritive
principles. While the liquor drinker's normal tissues are
absorbed and displaced by a hypernutrition of abnormal
deposits in the fibrous and areolar tissues, the opium
eater's adipose tissue is consumed and not replaced. There
appears, however, to be an inherent principle of limitation,
with respect to the amount of tissue thus consumed; as
the process stops when the surplus is absorbed, leaving
the system only sufficient to sustain itself.

The functions of the organs are performed under a constant protest. The internal organs of secretion are
contracted, the external are relaxed and moist, often
bathed in a cold, clammy perspiration ; the pupils of the
eyes are contracted, owing to a stimulation of the oculo-
motor centres. The blunted state of the gastric functions renders the assimilative processes torpid ; the brain
suffers for the want of nutrition and reflexes its in-

fluence by generating a limited supply of force to the nerves' periphery, inducing undue susceptibility to the action of cold and inability to sustain the shock of a cold bath. Loss of memory and dull perceptions invariably follow as legitimate effects of the drug ; dates, names, and past events become confused. The lips have a death-like whiteness, while dark circles appear under the eyes. Pruitis of the nose and lips is often a troublesome symptom. A remarkable effect of the drug is seen in its power to arrest diseases of a certain character, keeping them in a quiescent state, not by curative action, but by suspending their activity. This is exemplified in cancers of the breast or uterus, large and continued doses of opium arresting their growth, and, if they are discharging, drying their secretions. This condition lasts while the system is kept fully under the influence of the drug, but the disease resumes its activity with redoubled energy when the opium is withdrawn. Hemorrhage, in phthisis, and the progressive waste of lung tissue are controlled. But the full effects are procured at the expense of forming a habit. If the pains in rheumatism, nenralgia, or sciatica, are controlled by opium, they will return for a time on its withdrawal.

Renewed activity of all the functions follows a complete cure of the opium habit. The glands pour out their secretions. An excessive degree of vitality rapidly follows. The normal weight is soon reached and increased. Seminal emissions occur as a complication. Old gentlemen of seventy will suffer from that cause, their sexual vigor being as great as at any previous period of their lives, and continuing for a greater length of time than would be expected under the circumstances. It appears strange and unreasonable, and is not consistent with well-known laws of cause and effect, yet it has been proved under our observation to be a fact, that after restoration, no complications present themselves, indicative of permanent lesions, resulting from the habit. The functions are vigorous and active, seeming to have a new lease of life. The finer sensibilities return with full force ; desires and ambitions are awakened. If any lasting or obscure complications are superinduced by the habit, which develop in after years, and shorten life or undermine health, predisposing their subject to fall easily into disease, no such results have, as yet, indicated their presence by any premonitory signs, however slight. We wish to be fully understood as referring to patients who have been fully and radically cured.

During the first and second years of the habit, the system is not apt to desire or crave other stimulants than the opium. But finally the system becomes so prostrated and weak, that the opium is increased ; the fainting attacks become frequent, with palpitation of the heart, and vertigo ; the subjects of the habit feel compelled to resort to a diffusable acoholic stimulant with which to support the system and counteract distressing symptoms. Liquor is taken under a delusive idea of self-defense. Opium eaters in this way allay, temporarily, symptoms of weakness and vertigo, committing, however, a fatal error, by adding fuel to a flame that is consuming them, as liquor and opium combined induce fatty degeneration of the kidneys, which proves fatal within a few years. The opium depresses but leaves the system with a good foundation to build upon after stopping its use. The liquor counteracts and antagonizes the effects of opium, making an increase necessary to compensate for the deficiency. In their desperate efforts to sustain the system with opium, on the one hand, and to allay dangerous symptoms of narcotism by liquor, on the other, victims of the habit greatly increase the amount taken of both stimulants, creating deplorable complications. Their mental capabilities fail rapidly ; they become hypochondriac, and slowly descend to the lowest depths of degradation.

The alcohol consumed by them does not act as the hydro-carbons usually do, repairing their impoverished tissue it subjects them instead to its dilapidating influence, without bestowing upon them any compensating effects. Fortunately, only a small proportion of opium eaters allow the desire for drink to engulf them. By controlling the tendency to increase their dose, they enjoy comparative immunity from distressing symptoms, which require alcohol to make them bearable.

One of our patients, a gentleman of character and ability, succeeded in confining his dose by a daily allowance, to seven grains of morphia, making no increase for nineteen years. He forcibly described his trying position, by declaring it to be by a continual hand to hand battle with the monster, that he was enabled to ward off the fatal inclination to increase the dose. His mind and finer sensibilities were apparently unimpaired, a fact which makes his case a remarkable one, as it requires will-power and stamina to maintain such a course.

When opium and liquor are combined, and their quan-tity is gradually increased, enormous amounts can be consumed. A lady in our practice took each week four drachms of morphia with one gallon Tr. Valerian. A lady patient, married, thirty-five years old, consumed every month two ounces of morphia with five gallons of whisky, this quantity being allowed by her husband. She was not satisfied, however, and would pawn jewels, clothing and furniture in order to get more, so that it is impossible to determine the exact amount taken. The husband said "that he was in constant communication with pawn-brokers, having spent thousands of dollars in redeeming articles." The patient was a lady of rare attainments, belonging to a good family, having enjoyed all the advantages that wealth and a fine social position can give ; she formed a taste for wine at her father's board, and the opium habit during subsequent illness. Early impressions left her in a position to fall easily into excesses.

CHAPTER II.

To enable the physician fully to understand and appreciate the essential characteristics of the disease, as presented under a multiplicity of forms and circumstances, and to assure a perfect success, we give the record of individual cases, their history, indications, and treatment, with results attained. We also make suggestions as to changes and improvements, found by experience to be necessary.

The first great cause or incentive to take the drug, is pain in its varied forms. We have had presented for treatment cases differing in every essential particular, possessing little in common with each other. As the exciting causes are many, so the indications and complications, and temperamental idiosyncrasies play an important part, and materially assist or prove obstacles to the physician's work of treatment.

Ten patients take the same amount of opium. Its effects and power to control are dissimilar. Each one has his particular method and time for taking the drug that exerts its influence over the system. They consume different compounds of the drug. One uses Magendie's solution of morphia, hypodermically. Others take morphia by the stomach, or rectum, or by snuffing

15

it up the nostrils. Others use gum opium, laudanum, paregoric, dovers powders, McMunn's elixir of opium, soothing syrup, cough syrups and patent anodynes. Morphia enjoys the greatest popularity, as it is concentrated, leaving fewer bad after-effects. The hypodermic method is resorted to by many persons, and its results are more distressing than are those produced when the drug is taken by the stomach. As the habit advances they are obliged to resort to the use of the syringe from three to five times a day, filling the arms and hips with pertusions that excite hyperæsthesia of the cutaneous surface, and inflicting great pain by the operation. The arms become one mass of cicatrixs, resulting from abcesses and perforations, that leave indurated lumps under the cuticle, resembling to the touch a No. 2 shot. Finally, it becomes difficult to find a place large enough, between the nodules, to admit a needle ; the hips are used and soon present the same condition. The parts receiving a limited supply of nutrition, abcesses become indolent, and are slow to heal, often exciting atrophy of the muscles to follow.

If the patient is taking any of the compounds of opium, the physician should find by a mathematical calculation the equivalent of his dose in morphia and take the latter as a basis. Of gum opium it requires seven grains to make one grain of morphia. One fluid ounce of laudanum is equal to four grains of morphia. Dovers powder contains one grain of gum opium to ten grains of the powder. We have found that three grains of morphia would sustain a patient in the habit of consuming one bottle of McMunn's elixir of opium per day. Paregoric contains one and one-half grains of gum opium to the fluid ounce.

If the patient is addicted to any patent anodynes, cough or soothing syrups, the physician should find, by cautious experiments with minute doses of morphia, what amount the dose would represent in morphia. The painful disease for which the drug was first taken, should be carefully considered, as the opium exerts a peculiar power to suppress and hold in abeyance certain diseases. Their liability to return must be considered with reference to the result which their return would have upon treatment. Again, the suppressed disease may come to need treatment, and it must be asked in advance what success is reasonably to be expected in such an event.

Uterine and ovarian complications cause more ladies to fall into the habit, than all other diseases combined, and are also a prolific cause of trouble and vexation during treatment. The uterus being liberally supplied with nerves, vessels and glands, relaxation, for the want of opium, causes great pain, reflexing a decided action upon the general system. Women with child abort at once, and the miscarriage is followed by a uterine hemorrhage, frightful to contemplate, which requires promptness and decision on the part of the physician to ward off a fatal termination. The utero-placental vessels have opened wide their flood-gates, being atonic, and relaxed for the want of their accustomed support derived from the opium. The menstrual functions, when no pregnancy exists, are suppressed for months, indicating the profound effect exerted over the uterine plexus, by the habitual use of opium. When a cure is effected, the flow is profuse, accompanied with bearing down and expulsive pains; leucorrhœa is invariably present, whether the patient has suffered similar symptoms before, or not.

During the first few days after a cure is effected, the secretions are all profuse; the erectile tissue is acutely sensitive, simulating nymphomania. Such results, excited by functional disorders, enable one to form an intelligent conception of the troublesome complications to be anticipated, when a diseased condition of these parts really exists. If a patient has taken opium for a uterine disease, she ignorantly assures the physicians, in many instances, of its supposed cure; basing the assumption on the fact that the symptoms have not been present while the drug has been in use. The physician must not let this declaration mislead him, but should prepare to render prompt, well-directed assistance during the crisis. Many patients feel ashamed of being addicted to the drug, and wishing to retain respect, will tell the physician of some imaginary trouble as the cause of the habit. He should accept their statement and give them his sympathy; passing judgment upon the case, however, according to the indications.

Diseases acting spasmodically, and ataxic symptoms, must be met and treated as if occurring under any other circumstances. There is a period of twelve hours only requiring active interference with aggravated cases of the opium habit, in which there is danger of a return to the drug, through the severity of the patient's sufferings. It is necessary to be on the alert during that time, and to bridge over the critical period, by direct efforts to control spasmodic nervous action.

A serious question arises when young married ladies apply for treatment and declare that they are with child, not having menstruated for five, six or eight months. This condition is natural to the opium habit; yet the patient may be enceinte, and the physician must form a correct opinion. If the woman is with child, an immediate cure is out of the question. Treatment must be postponed until it will not imperil mother or child, as is suggested under the proper head. If there is suppression only, you do not wish to lose time. Question the patient categorically, and find out what her habit has been heretofore; further than this you can rely upon the usual signs in pregnancy; they are not otherwise modified. If not fully convinced, and entertaining reasonable doubts, avail yourself of the judge's right, which is also yours, and reserve your decision. No case is too old for treatment; the patient who has taken the drug for twenty years is as susceptible to treatment (other conditions being favorable) as one who has used it during three years only. The amount consumed does not stand in the way of a radical cure. We call to mind an easily attained victory over the drug, where it had been taken for twenty years, and where, during the last nine years, one drachm of morphia was consumed daily.

By carefully pursuing the methods suggested, noting special features connected with the following cases, a clear conception may be had of the principles involved in the successful treatment of the opium habit.

CHAPTER III.

Mrs. B. L., married, thirty-two years old; general health, previous to forming the habit, good; when treatment was begun she was consuming fifteen grains of morphia per day. The incentives prompting her to the indulgence were worry and suspense concerning the uncertain fate of her husband, who was in the army. A neighbor, and without doubt an opium eater, suggested the use of the drug as a means of drowning trouble. After taking morphia for six months, constitutional weakness and prostration prompted the victim to make a desperate effort to throw off the habit. After suffering the torture of relaxation for fifteen hours, she abandoned the effort, unable to endure the pain, and resigned herself hopelessly to the practice.

The power of opium to stimulate and soothe was lost to her; yet she was obliged to take her usual dose daily. The history of a subsequent attempt, made by the woman's husband, to break the habit, is full of interest to the physician, illustrating as it does the power exercised by the drug over the system. On his return from the army, he learned the facts connected with his wife's habit for the first time and determined to take matters

21

into his own hands and effectually to cure her. Calling a sister to his aid, he unfolded a brilliant plan of procedure. This sister heartily seconded his resolution, promising her help and co-operation, the more readily because she had been often humiliated by her sister's oddities, which attracted attention and prompted unfavorable comments. The victim of the scheme expostulated, and endeavored to impress upon her husband and sister the impossibility of carrying out their plan; but remonstrance was to no purpose, as her tormentors belonged to that unfortunate class of persons who can learn only from that severe, but ingenious teacher, experience.

The family lived in a lonely and isolated farm-house on the prairie, between two villages, one six, the other four miles distant. It was a two-story frame building, with a veranda in front. The plan adopted, which met with an unlooked-for termination, consisted in a vigorous display of brute force. The husband destroyed his wife's supply of opium and made her a prisoner in an upper front room, the pair of amateur physicians relieving each other in a constant watch upon their patient. In locking her up, they thoughtfully withheld all clothing except a night garment.

Stopping the supply of opium in the morning, they prepared for a long siege. The patient went through the day with comparative ease, entertaining a hope that the trial would not be severe. With all her fears she was happy in the thought of a possible cure. As night approached, these hopes were dissipated and an awful dread of impending danger took their place.   The slumbering forces within her seemed, for a time, to be quiescent, leading her to imagine that the critical period had passed ; but, as with a volcano preparing for a fresh eruption, this proved to be the quiet preceding upheaval. Premonitory symptoms, indicative of the crisis, gave her a foretaste of the exquisite horror to be endured. The nerves seemed to be crawling beneath the flesh, their serpentine motion compelling her to keep in constant action.   Her mental faculties were unimpaired, her judgment being clear and unclouded, and the phenomena presented being purely of a physical nature.

At one moment her blood seemed to be on fire, and this was succeeded by the coldness of death striking into the very marrow.   A sensation at the pit of the stomach was suggestive of the possibility that the vitals were in process of destruction.   Spasmodic vomitting attacks prostrated her with their severity, leaving the surface of her body bathed in a cold, clammy perspiration ; the bowels moved every ten or fifteen minutes, the motions being accompanied by burning tenesmus, and followed by expulsive and bearing down pains ; her catamenial discharge appeared with the profuseness of a hemorrhage, this being the first time in nine months that she had menstruated.   Her sufferings were now agonizing in the extreme ; gaping, sneezing and stretching, and a quick incessant cough, excited by relaxation of the uvula, caused symptoms of suffocation.   She begged her friends, if they had pity, to kill her at once and not to prolong her misery. She said afterwards : " If no way of escape had presented itself, I would have dashed my brains out on the stove."

Her muscles twitched and contracted with such intensity as to throw her from chairs. An excessive flow of tears and saliva followed. She believed confidently that her sufferings would drive her insane before morning, as her family would not, or could not, realize her critical condition, and gave no heed to her pitiful appeals for help.

During the day a cold November sleet set in, driven with merciless force before a sharp north-western gale ; its velocity unimpeded in its course across the prairie, made the night dark and foreboding, rendering the road almost impassable. The suffering wife, alone, at the mercy of an ignorant, determined majority, that could not appreciate her terrible condition, with the evidence before them of nervous prostration and relaxation, controlled herself to the best of her ability, until one o'clock, when her husband retired, leaving her sister to continue the night watch.

In making the change, both were out of the room for a moment. The woman made use of the opportunity by emptying the drinking water from a pitcher into the chamber vessel. Giving her husband ample time to become comfortably settled, she requested a drink. While the sister was obtaining a fresh supply, the poor, misjudged wife, weak from prolonged misery, with the curse of woman upon her, leaped from the window to the veranda, letting herself down to the ground. Like the slave in the dismal swamp, she was free, but with fearful surroundings. The sleet froze as it touched her only garment ; but the opium horror was upon her, and she started across the lone prairie, sinking ankle deep in the freezing mud at every step. If mad, "there was method in her madness," as she started for the village six, instead of the one four miles distant, believing that pursuit would be instituted at once in the latter direction.

She anticipated the movements of her watchers correctly. Her husband, responding to the sister's frightened cries of alarm, searched the out-houses and started for the nearest village. The wife walked the lonely country road, facing the blinding sleet, her only garment heavy with ice. Arriving at the village, she awoke the druggist, a sensible gentleman and physician, who fully realized the terrible necessity that would drive a lady to such extremes, her disposition being modest and retiring ; naturally possessing none of the characteristics likely to lead her to undertake a journey over the wild western prairie, clad in a night garment.

After obtaining what her system demanded, she was obliged to wait but a short time before her husband arrived, who, being thoroughly convinced that her craving was not an appetite indulged for the sake of pleasing effects, purchased a supply of opium, and returned home a wiser man.

In forming the habit, this woman had taken one-eighth

to one-fourth of a grain of morphia twice or three times a week, increasing the amount progressively, until fifteen grains per day were consumed. An only child, born four months subsequent to the formation of the habit, resorted early to the bottle, as the secretion of breast-milk was not sufficient to sustain life. The same jerking and twitching of the muscles that affected the mother, was noticeable in the child. It was frail and weak, with relaxed bowels, sleeping but little and cry-ing incessantly. It is owing to the gradual with-drawal of breast-milk and the early resort to the bottle, through a suppression of lactescent activity that the lives of many children, born to mothers in the opium habit, are saved. When the woman was brought under our observation, we found that she was suffering all the decided effects of the opium, being sallow and anæmic, displaying the opium cachexia, and weighing less than a hundred pounds, her normal weight being 135 pounds.

She was in the fifth year of the habit, menstruating once every seven months, while her habit, before taking the opium, had been to flow freely for six days each month. Constipation, with painful hemorrhoids, also anorexia and palpitation of the heart, aggravated her case. She could not sew or read without going to sleep. A lamp was kept burning during the night in her sleep-ing room. We find this desire for a bright light to be universal upon the part of opium eaters. The friends of this patient declared her to have been a bright, ambitious woman, before forming the habit; now, but few evidences remained, indicative of her pristine state, she being dull and stupid, and afflicted with strange oddities.

Under the influence of a full dose of morphia, she could, however, do a great deal of manual labor; as it stimulated her to unnatural effort, only to suffer vital reaction, and fall away still further from a normal standard. We considered her case free from complications, and so informed her husband.

It is injudicious to go into particulars regarding therapeutical methods, as it is sufficient for patients to know that the opium will be withdrawn and the system sustained, if they follow instructions implicitly. Our directions were minutely gone over with the husband and sister, to impress upon them the importance of system, with favorable hygienic measures.

Experience has demonstrated the value of customary surroundings, for patients while under treatment. We have treated opium eaters and dipsomaniacs in asylums, boarding houses, and large hotels; and two of our patients kept travelling constantly, in order to derive the benefits incident to a change of scene, and also to occupy time, and so divert their thoughts from themselves. The best results are attained at home, however. We instructed the patient now under consideration, to pursue the even tenor of her way, attending to her usual duties, being particular not to excite fatigue by overwork, and to avoid excesses of all kinds; as over-exertion would excite reaction, demanding increased stimulation, which, if indulged in, destroys the systematic course of treatment.

Patients consuming one drachm of morphia per day
feel no sedative effect in proportion to the amount taken.
The system cannot assimilate or appropriate more than
a given amount; and any excess tends to produce
dangerous prostration.   The physician should endeavor
to estimate the minimum dose necessary to sustain the
system, and compound the patient's medicine on that basis.

As the patient in case No. 1 was consuming fifteen
grains of morphia per day, we ordered her medicine compounded on a basis of ten grains per day, for sixty days,
in the following manner:

There were two bottles, branded respectively No. 1 and
No. 2.   No. 1 contained the medicine to be taken as
directed; No. 2, a supply to be used in replacing that
which was taken from No. 1, as directed.   The bottle
No. 1 contained,

> ℞.   Morphia, ℈ x.
>   Alcohol, ℥ viii.
>   Gentian comp. Tr. ℥ x.
>   Ginger Tr. ℥ vii.
>   Aqua, ℥ xxv.

M. Sig. two teaspoonsful after each meal.

The amount taken from this bottle must be replaced
from No. 2, as directed.   Bottle No. 2 contained the
following:

> ℞.   English Extract, solid.
>   Cannabis Indica, ℈ v.
>   Glycerine, ℥ xx.
>   Alcohol, ℥ xviii.

M. Sig.   To be used in replacing that which is taken from No. 1.

The alcohol is necessary to dissolve the cannabis, and the glycerine avoids precipitation of the cannabis. As accuracy in the amount taken is absolutely essential, and teaspoons vary in size, a graduated glass should be used. THE PATIENT'S USUAL DOSE OF OPIUM MUST BE DISCONTINUED AT ONCE, AND NO REACTION WILL BE EXPERIENCED IN MAKING THE TRANSFER.   The change from opium, as contained in No. 1, to the cannabis, ginger, etc., in No. 2, is so gradually induced that the system receives no shock. The cannabis being a powerful tonic and antispasmodic, excites renewed activity of the gastric functions, stimulates the assimilative processes, equalizes the circulation, and combats and controls the characteristic spasm of the nerves, when relaxed for the want of opium, as no other drug known to us does.   Belladonna and cicuta materially assist in supporting the nerves during the crisis, but do not allay the peculiar state so dreaded by the victim of opium.   We find that solid English Extract is less apt to precipitate than fluid extracts, and fluid extracts of cannabis do not retain their active principle, with entire uniformity of strength.   If the mixture precipitates, the physician must recompound, adding gum arabic, making a heavy solution.   Our patient, as is invariably the case, experienced pleasing results from the change.

By the twentieth day of treatment, she had made wonderful progress, having a voracious appetite, with a gain of eleven pounds in weight. Stupid and

drowsy sensations being dissipated, a rest was now
ordered, to enable the system to adapt itself to the reduction attained. The use of bottle No. 2 was suspended
for nine days, and the patient was instructed in resuming its use to replace the amount taken from No. 1 only
every other day. Her husband was asked to report when
reaction set in. She was now consuming but a small
amount of opium per day, and the results were marked,
her eyes having resumed their old time brightness, while
she was happy in the thought of soon being free. An
exhilarating sense of relief is enjoyed when this much is
accomplished, giving patients a foretaste of the happiness
awaiting them when fully cured. This condition is
opportune, as it inspires them with hope and strength
to endure the pain incident to complicated cases during
the crisis.

Our patient presented no troublesome symptoms, her
bowels relaxing some days before the crisis. She had
two free actions per day, an important condition for the
physician to note, as it foreshadows a complicated or
uncomplicated crisis, rendering prompt attention necessary. Subsequently we will refer again to this important
symptom. On the fiftieth day of treatment, we found
that the patient had gained in every way ; she complained, however, of bearing down pains and leucorrhœa.
We ordered a sitz bath of warm water, followed by a
carbolized injection   The menstrual flow appeared the
same night, and we decided to make no further decrease
in the strength of mixture No. 1 during its continuance.

As nervous, ataxic symptoms are generally exaggerated for the time being, No. 1 was continued as usual, without reducing with No. 2. The patient was taking but an infinitesimal dose of morphia. Although she slept but little, she suffered no nervous irritability. The menstrual flow was profuse, lasting nine days. The use of mixture No. 1 was stopped the day following, and the patient was given phosphoric acid, ten drops—diluted—every half hour, until sixty drops, or one drachm, had been taken, allowing two hours to intervene before administering more. Caution must be exercised in the use of this acid, as gastric disturbances are excited if the medicine is given in excess. We compound it in the albumen of egg, as a vehicle affording more protection to the coats of the stomach than any other substance known to us.

Phosphoric acid is a powerful antispasmodic, yet many medical writers do not recognize the fact. Its sedative action in allaying nervous excitability, is pronounced as trust-worthy. A liability to induce gastritis, if its administration be prolonged, renders, what would otherwise be a master of opium, a partially useless weapon. Its use should be confined to a few hours subsequent to withdrawing all opium.

When the crisis is at its height, the physician must not allow his anxiety to control spasmodic nervous efforts to influence him to continue its use against his judgment. We exhausted our ingenuity in an endeavor to discover a vehicle that would protect the stomach, but

with only negative results. There is great efficacy in
the hot water salt bath. Indeed, it surpasses in its
tranquillizing effects all the narcotics—except the drug
opium.   With this class of patients, passing through the
crisis, it soothes, by equalizing the circulation, allaying
spasmodic activity of the nerves, and inviting sleep.   To
neglect or omit its frequent use, will bring about defeat
at the threshold of victory. At this period of treatment patients are prostrated by relaxation and very susceptible to outward impressions, having considerable
hyperæsthesia of the cutaneous surface.   They strenuously object to the bath.   The physician must respect
their trying position and explain its action, when his
advice will be followed, but the patients will have no
confidence in its good results.   Afterwards, being agreeably surprised by its soothing effects, they increase their
confidence and respect.   If the bowels are thoroughly
relaxed, it is well to inform the patient of his exact condition, as the time is favorable and the system free from
opium in any form ; the crisis will be over within
twelve hours.   Knowledge of this fact encourages the sufferer to endure his pains ; he will desire to repeat the
bath as often as possible to obtain its soothing effects.
It is safe to yield to this desire, as only good results can
follow.   Diluted phosphoric acid was administered to our
patient six times during the night ; she being greatly
improved by morning, the subsequent two nights were
passed with a moderate degree of comfort, as symptoms
of irritability were allayed by an occasional bath, making a

complete and rapid recovery.   In answer to our inquiry, two years afterwards, she reported herself well and fleshy, with no desire or inclination for opium, being doubly blessed by the presence of a twelve-months-old girl baby.

This is our simplest method for treating cases without complications, and is easy of application. We have relieved many patients in this manner, depending largely upon the baths. Complicated cases will, however, present themselves, when routine practice would not only be vicious, but also rob the physician of favorable results. Paralysis of the bowels will be met sometimes, with loss of vigor and tonicity upon the part of the general system, necessitating the tracing of effects to causes, and the application of appropriate remedies. Ingenuity will be required to relieve these symptoms.

CHAPTER IV.

CASE NO. 2. Mrs. Julia L., 31 years old, 5 years married. The incentive inducing her to take the drug, was association with a sister who was an opium eater.

She possessed a delicate organization, with hysterical tendencies, enjoying, however, apparently good health before forming the habit, although her immediate friends supposed her to be consumptive. Seeing her sister take the drug, she would occasionally indulge, and being frail and easily influenced, soon formed the habit. We will only refer to her sister's case, to note the fact of her cure within twenty-eight days, by the treatment used in case No. 1, except that no phosphoric acid was used, as there was no insomnia or irritability demanding its administration. The patient was thirty-five years old, and had been a confirmed opium eater for five years, consuming eighteen grains of morphia per day.

Patient No. 2 on coming under our observation, was consuming twelve grains of morphia per day. When she was fatigued by over-exertion, the dose was increased; the morphia supporting her during such emergencies, as the power to undergo physical endurance under its action is wonderful. While prostrating in the end, its direct effects are to sustain the system.

Our patient's natural tendencies rendered her susceptible to the pestiferous effects of the poison, so that she early fell under its influence and was reduced to a skeleton. In appearance her skin was dark and jaundiced, indicating a degeneration of the nutritive constituents of the blood; the hair and nails ceased to grow, the latter becoming brittle, showing a suspension of their nutrition. As is usual with opium eaters, anorexia and constipation aggravated her case. She had not menstruated since forming the habit, and had imagined herself to be with child for some months. During the tenth month of the practice, her family were horrified by her having a hemorrhage, apparently from the lungs. It did not suggest itself to them that the habit was the exciting cause of the suppressed menses and its vicarious elimination from the system, by hemorrhage. Her strength failed progressively from this time, the hemorrhages recurring, with some degree of regularity, every three or four months. She was given up as irrevocably doomed to slow consumption, a weak, hacking cough giving color to the supposition.

We considered her case a desperate one and so informed her family. She insisted, however, upon being treated, if only that she might die free from the monster, opium. In order to decrease her consumption of morphia slowly, we prescribed the same amount as was contained in the mixture compounded for patient No. 1, changing her No. 2 mixture in the following manner:

R.   Cannabis Indica, ℨ v.

        Belladonna Tr. ℥ vi.

           Glycerine, ℥ xv.

             Alcohol, ℥ xx.

M. Sig. As she used No. 1, replace from this bottle, *only every other day.*

Salt Baths were ordered to be taken three times a week; the diet to include a liberal allowance of fruit and vegetables and a lemon or orange was ordered to be taken before breakfast and on retiring. If the bowels in these cases do not respond to a fruit diet, it is necessary to facilitate their action every other day by an enema, consisting of one ounce of castor oil. As there was general poverty of the nerve centres in this case, we ordered syrup of hypophosphites, taken alternately every other week, with the following:

R.   Iodide Lime, gr. x.
Phosphate Iron, 3 i.
Quinia, 3 i.
Lactopeptine, 3 ii.
Syrup simple, 3 v.

M. Sig. Teaspoonful at nine, three and nine o'clock.

During the subsequent forty days this patient's improvement was phenomenal, and was accompanied by a ravenous appetite. She gained flesh at the rate of three pounds per week. Her bowels did not, however, relax, *or show any disposition to regulate themselves, displaying an atonic condition, which it was absolutely necessary to overcome before a cure could be effected.* On the thirty-fifth day of treatment she had a hemorrhage, more profuse than usual, succeeded by hemoptysis for three days.

The lime, iron and quinia were discontinued, and the following pill was given:—

R.   Ferri sul. gr. xv.
Colocynth, ext. gr. x.
Henbane, ext. gr. iv.
Leptandrin, gr. iii.
Podophyllin, gr. ii.
Aloes, gr. iv.
Capsicum, gr. v.
M. Pills xxv. Sig. One pill after meals.

Some years previous to forming the habit, the patient had suffered dysmenorrhœa and leucorrhœa, receiving treatment at that time for ulceration of the os-uteri. An examination displayed a congested and thickened os, with two or three cicatrixes, the results of former ulceration. On the seventieth day of treatment, she experienced for the first time expulsive pains, severe in character, accompanied with backache and followed by leucorrhœa. Warm injections of castile soap water, preceded an injection of tea twice the strength of that commonly used at the table, and as warm as was consistent with comfort. The next morning we ordered the castile soap water repeated, using the following as a final vaginal injection:

R.   Glycerine, ℥ iii.
Carbolic acid, ℨ ii.
Camphor aqua, ℨ i.
Aqua, ℥ x.

M. This, in a measure, controlled the symptoms, but we were hastily called three days afterwards, and found the patient suffering general prostration. The bowels had not acted for three days, the movements excited by injections were unsatisfactory, giving no relief. Anorexia being complete, the sight or smell of food induced nausea.

With our present experience we would not pursue the course resorted to in her case, where *the bowels were unrelaxed.* As it was, the prescriptions Nos. 1 and 2 were stopped and baths ordered. Electricity was applied with sponges over the abdominal viscera and rectum, exciting a passage, which was, however, scant, and forced, and not sufficient to relieve the system. Calomel of the tenth trituration, with full doses of podophyllin, was administered during the evening. At four o'clock the following morning, we were called and informed by the messenger that our patient was dead, having breathed her last a few moments before. She was indeed dead to all appearances, being in hysterical catalepsy, with no appreciable action of the heart or respiratory muscles. She had suffered greatly during the night, vomiting incessantly, with no action upon the part of the bowels. We administered, hypodermically, one-half grain of morphia, when a little cold water sprinkled in the face excited reflex centric spinal action and revived her. *This instance only confirmed the conviction that it is impossible to cure the opium habit, and bridge the patient over the crisis, without having the bowels freely relaxed.*

The condition unmistakably indicates,—and the indication should not be misinterpreted,—a state of the nerves'
periphery, which affects the system at large by a reflex
action, showing that nature is oppressed by some obstacle
which precludes the possibility of an immediate cure.
The indications are broadly presented, demanding that
no further effort be made to reduce the dose. The
patient should be put on the smallest amount of opium
consistent with a quiescent state of the nerves, and means
should be taken to build up the general health by the
judicious administration of tonics, to excite deposits of
nutritive principles that give tone and strength to the
nervous system.

A rule, scrupulously to be observed, is *not to allow the
patient to advance into the crisis until the bowels have
freely relaxed, involving the entire canal.* The crisis is
a condition following the withdrawal of the last infinitesimal amount of opium. In preparation for it, patients
may be kept as near the verge as the physician wishes,
and they will improve, it being only a question of time
when their improvement will revivify the antonic nerves.
The activity of the nerves' periphery, presiding over the
abdominal viscery, will be a true criterion of their condition throughout the system and a signal for the treatment to be resumed in safety, with victory near at hand.
Drastic cathartics will not facilitate the action of the
bowels, as paralyzed nerves recognize no such master.

We kept our patient on a small quantity of opium,
slowly reducing that amount every third day, allowing

tbe system time to recuperate. We prescribed the following:

  ℞.   Morphia, 3 ii.
         Alcohol, ℥ v.
        Glycerine, ℥ vi.
         Aqua, ℥ vii

M, Sig. Teaspoonful after meals.

Bottle No. 2 contained :

  ℞.   Cannabis Indica, 3 vi.
       Belladonna Tr. ℥ iii.
        Alcohol, ℥ iv.
        Ginger Tr. ℥ v.
     Gentian comp. Tr. ℥ vi.
    Syrup Ferri Iodide, ℥ iv.

M. sig. Every third day replace what is taken from No. 1, with the above.

We directed the patient's husband to inform us at once when her bowels fully relaxed. Thirty-seven days afterwards our presence was requested ; we found her greatly improved in every respect, presenting quite a natural appearance, her bowels having relaxed the previous night, moving twelve times before morning, with accompanying expulsive pains and profuse vaginal secretions, her catamenia appearing for the first time in three years. The attendants kept the first large discharge for our inspection, as it excited their curiosity by its peculiarity of character. It consisted of a mass of black coagulated matter, thickly studded with fibrinous laminæ, or flakes, emitting a putrid odor ; also a mass of remarkably hard scybala, having stamped on their surface the imprint of numerous crescentic folds from the columnar epithelium, showing that it must have remained impact in one spot for some time. The relief experienced by the patient was complete, although she was exhausted. Prescriptions Nos. 1 and 2 were stopped and the patient was given one grain of quinia every hour, with instructions to chew coca leaves, retaining the juice extracted, which enabled her to pass safely through the crisis, without suffering nervous irritability. Within five days she was doing housework.

A letter from her brother, who is also a physician, written two years later, gives a glowing account of her perfect health, hemorrhages and other phthisical symptoms having disappeared, menstrual functions being normal, while her former frail state was entirely gone and replaced by robust health.

CHAPTER V.

CASE No. 3. Mr. H. L. C., commission merchant, 39 years old. He enjoyed good health previous to his 31st year, when a catarrh, mild to that date, assumed a serious form, exciting the usual train of symptoms characteristic of the disease.

On his coming under our observation, we found him afflicted with chronic catarrh accompanied by bronchitis ; the voice indicating a chronic coryxa. There was also an inflammation of the pharynx and larynx, involving the large bronchial tubes. He experienced a sense of pain and constriction beneath the sternum, and complained of frontal headache and lassitude, his appetite being poor, and his state one of general malaise. The opium suppressed these troubles and controlled their activity. When the system was not fully under its direct influence, the secretion of mucous was profuse, and had an offensive odor. After experimenting with different methods of treatment for the cure of his catarrh with but negative results, he was so unfortunate as to meet with one of those indefatigable beings (and they are generally conveniently at hand) who give advice gratuitously, and display their profound knowl edge of a science they have never studied. Complications of every character are easily recognized by them and the treatment necessary is confidently prescribed, when men expert in diagnosis, who have devoted their lives to its study, are uncertain of the origin and nature of the trouble. These street corner advisers quickly solve the problem. Such a one told our patient to dissolve morphia in water and snuff it up his nose, promising a radical cure as the result. The relief afforded was instantaneous, as the morphia stopped the secretions, suppressing the flow of mucous, working in this manner its full physiological effects as completely as if it had been taken hypodermically. The usual prostrating results followed.

The patient attempted to stop its use, after discovering the fact that the opium was slowly dementing him. Do what he would, it held him bound hand and foot. When his case was brought under observation he was in the fifth year of the habit, and was consuming thirty grains of morphia per day, by making a saturated solution and snuffing it through the nostrils, losing but a small percentage of the drug, as it was absorbed rapidly. Although his case was complicated, we determined to give him a vigorous treatment, as he was anxious for relief, and declared that his mind was slowly leaving him. His actions and appearance strengthened this supposition. His eyes were sunken and expressionless. While describing his symptoms, he fell into a doze, waking up with a violent start. His opium dreams were horrible beyond description. For hours he would lie as if transfixed, unable to control a muscle, bathed in a cold perspiration, while slimy leeches fastened themselves to his lips. Mouldering bones and sluffing flesh were piled upon him. He made desperate attempts to free himself, consuming, as he thought, days and weeks in the effort. On waking, he would find to his surprise that only a few minutes had been passed in sleep.

The scenes enacted in his dreams were so vividly terrible as to make him look with horror upon the approach of night. Weak and prostrated, suffering for the want of natural rest, he dared not indulge the inclination to sleep until nature forced him to yield. He would also be attacked by sinking sensations, as if falling into a pit, on going to sleep ; and it required a strenuous effort on his part to arouse himself. He sought relief by resting in a chair, but to no purpose. We recognized the necessity of giving battle to the catarrh, which would otherwise stand as an insurmountable obstacle to success, and ordered him to stop snuffing the morphia, and compounded mixture No. 1 on a basis of ten grains of morphia per day, in the following manner :

Bottle No. 1,

    ℞.   Morphia, 3 x.

                 Alcohol, ℥ v.

             Glycerine, ℥ x.

               Aqua, ℥ xxv.

M. Sig.   Two teaspoonsful after each meal.

Bottle No. 2 contained :

R. Cannabis Indica, 3 vi.
Belladonna Tr. ℥ x.
Gentian comp. Tr. ℥ xii.
Ginger Tr. ℥ viii.
Glycerine, ℥ x.

M. Sig.   As the dose is taken out of No. 1, replace it from No. 2.

For the catarrh we ordered :

R. Chlorinated lime, ℥ i.
Aqua, ℥ vi.

M. Sig.   Inject into the nostrils, so as to reach the mucous coats covering the posterior narès and inferior meatns, clearing the parts thoroughly.

To be followed in one hour by :

R. Carbolic acid, 3 i.
Glycerine, ℥ i.
Aqua, ℥ iv.

M. Sig.   Inject as before.   If the parts feel sore or raw, blow through a tube the following :

R. Chlorate Potash, ℥ i.
Pulvis Peruvian, 3 iii.

M. Two or three times each day.

We also had him inhale the steam of Tr. Iodine.   This so loosened the deposit, as to enable him to throw off an accumulation of putrid secretion.   The parts bleeding somewhat, we used at once carbolic acid and glycerine by injection, affording immediate relief.

After getting the upperhand of the catarrh we discontinued the injections, as we do not consider it good practice to inject liquids to any considerable extent, into the nasal cavity and pharynx. To do so affects the eustachian tube, causing deafness in many cases. On the third day we found that mixture No. 1 was not strong enough to sustain the man, and added one drachm of morphia. After the seventh day of treatment the patient began to improve, his appetite responding to the tonic, while its action, by equalizing the circulation, enabled him to sleep naturally. As a constitutional treatment for the catarrh we prescribed:

R.   Iod. Pot. ʒ vi.
            Tr. Iodine, ʒ iv.
        Bi. Chlo. Hyd. gr. i.
            Aqua, ℥ vii.

M. Sig. Teaspoonful in a wine-glass of water after meals.

The patient now improved rapidly; on the twenty-fifth day of treatment he had gained eight pounds, was appearing well, and sleeping quietly. We judged, by careful questioning, that it would not be safe to attempt a further reduction of the morphia until the system had adapted itself to the amount now taken. The symptoms shown in his case, insignificant in themselves, when connected with the opium habit, are indications which should be recognized. He was easily fatigued, hot and cold flashes were experienced before the hour arrived at which to take his regular dose, and perspiration was easily excited, accompanied by gaping and sneezing.

We ordered him to stop reducing mixture No. 1 with No. 2 for ten days. In overcoming the results of opium, procrastination in treatment robs you of good results. Certain conditions must be met and subdued as expeditiously as circumstances will allow, in order to obtain favorable results and lasting good. The remedies used are powerful, and in order to reap the benefits of their fine medicinal properties, the physician should attain the desired end before their use becomes second nature to the system. At the same time, he must avoid what would be calamitous, namely, the crowding of the system beyond its ability to ward off a shock that would imperil the desired results of treatment. If a mistake is made a certain amount of confidence is irrevocably lost; the patient's first and strongest determination, like a broken rope, can be nicely mended by an experienced hand, but its elasticity and continuity are gone. The nervous system, like a stringed instrument, is susceptible of being cautiously wound up to a high degree of tension; but, if manipulated carelessly, it suddenly breaks at one-half the strain. When this accident once happens the patient is apt to look anxiously forward to a repetition of the experience.

During the ten days of rest provided for the system, by the suspension of mixture No. 2, the patient improved;

all symptoms indicative of approaching relaxation disappeared.   We ordered bottle No. 1 refilled by putting in
two drachms of morphia; and added to No. 2, Cannabis,
six drachms, filling with Gentian, Belladonna and Ginger.
Considering this amount sufficient to complete the cure,
we directed the patient to replace what was taken from
No. 1 every other day, thus allowing the system ample
time to adapt itself to the change.

During the subsequent thirty days he made rapid progress towards an improved condition; the bowels regulating themselves, his appetite and assimilative processes
acted harmoniously, enabling him to gain two pounds
in weight per week.   All this so encouraged him that
he joyfully notified us of his complete cure; he could
not appreciate the necessity for taking more medicine.
But it is necessary to resist this tendency on the part of
patients to stop treatment.   They feel so well that they
do not comprehend the force and power of the last link
in the chain that holds them.   We warned this patient
of the disastrous consequences that would inevitably
follow a sudden withdrawal of his medicine; assuring
him that in that case he would become so nervous in a
few hours as to require a large increase of opium.

Five days afterwards he reported an inability to sleep
after one o'clock at night, which created discomfort
bordering on irritability.   A hop pillow, as a mild
soporific, was used, sprinkled with reduced alcohol to
avoid crackling and to bring out its aroma.   On the
seventy-eighth day of treatment he sent for us.   We

found him fully in the crisis; his bowels had moved every few minutes, during the last three hours, and he was wakeful and irritable. Mixture No. 1 was stopped. When an attempt was made to administer fifteen drops of phosphoric acid, vomiting set in. A warm salt-water bath quieted the nervous symptoms, but did not allay the vomiting. A mustard plaster was applied over the epigastric region, and ten drops of chloroform, ad- ministered with the white of an egg, controlled the irri- tability of the œsophageal and gastric branches of the pneumogastric nerve, stopping the vomiting and enabling the patient to retain the phosphoric acid.

The acid was followed by five-grain doses of musk every half-hour, with warm baths every hour. The patient passed the day in comparative comfort. The secretion of mucous from the nasal cavity being profuse and continuous, the treatment prescribed for that condi- tion in the beginning was resumed, but failed to make any impression. The seminiferous secretions were active, involving frequent seminal emissions. General prostra- tion, anorexia, sneezing, gaping, with occasional vomit- ing, constituted the principal symptoms calling for our interference. These symptoms are generally present during the crisis, with patients laboring under wasting diseases. The nervous manifestations are of a decided character, and suggest intense suffering; when such is not the case they denote absence of equilibrium upon the part of the nervous forces, excited by relaxation of the central functions, which allows spasmodic effort to go unchecked. The flood-gates of the system are wide open, after being unnaturally closed by the opium, but fortunately this condition does not inflict upon the sufferer that unendurable agony, not appreciable to the eye of an observer which follows the sudden withdrawal of the opium.

That evening, musk in ten-grain doses was administered to our patient and by having him inhale, one hour afterwards, the fumes of nitre paper, six hours' sleep were enjoyed. Nitre paper will often induce sleep when musk, cannabis, lupuline and lactucarium fail. Each movement of the bowels excited painful tenesmus. We adopted Dr T F. Cock's plan for relieving this, with marked success. This is done by putting a drachm of chloroform in the bulb of a syringe, screwing the tube on, (taking care to keep it upright, so that the chloroform may not get on the outside, where by any chance it might touch the inflamed parts,) and inserting the tube in the rectum, allowing the chloroform to evaporate. No danger is incurred, and prompt relief follows. Under the combined influence of phosphoric acid, musk, nitre paper and baths, our patient passed a quiet night, his appetite returning the next day. Extreme prostration, but no irritability, was present.

He described his feelings by declaring himself to be contented and satisfied, with no disposition to exert himself in any way. His condition was one of nervous relaxation and prostration without irritability. His bowels were still relaxed. We generally allow the latter to regulate themselves, which they will do in from four to six weeks. To counteract the languid, tired feeling, we ordered coca leaves, used as recommended for case No. 2. The patient secured only about five hours' sleep each night, the wakefulness, however, not being unpleasant. The emissions recurred as he fell asleep. As with relaxed bowels we do not attempt to correct that condition, as it disappears when the prostration attending central relaxation subsides. The catarrh was troublesome through an excessive secretion; it retained, however, none of the characteristic signs indicative of chronic bronchitis, and we felt confident that the secretion would stop with the general equalization of the emunctory functions. The catarrh treatment, both local and constitutional, was, however, continued.

An encouraging assurance which the opium eater has during treatment, is that no relapse is likely to occur. The system maintains the position gained, unless chloral, bromide of potassium, or liquor is taken, either of which would further excite peripheral relaxation. We shall treat of that state under its proper head.

Eight weeks afterward, this patient reported favorably. He was gaining flesh rapidly, and weighed seven pounds more than ever before. His weakness was entirely gone, his sexual desire was still abnormally strong. The catarrh was greatly improved, showing a tendency towards a complete cure. A remarkable result, observed in all cases, exemplifying the fact that stimulation excites a desire for increased stimulation,

was noticed in this case. The patient had been an inveterate smoker. Previous to forming the opium habit, he
indulged incessantly ; as his treatment for the habit
advanced he cared less for his pipe. During the crisis
he entirely stopped smoking, returning to business without thinking of, or desiring tobacco.

In a report made two years later, he assured us of his
continued good health and happiness, but blamed himself for having resumed the use of tobacco. After he
had gone five months without smoking, he had returned
to the practice in a social way, and not because of any
felt necessity of the system. He had been free from
catarrh for several months.

CHAPTER VI.

CASE No. 4. Captain C. B. H., 39 years old, a lawyer by profession, formed the habit through a painful wound, which prostrated him for some months, resulting in ankylosis of the knee joint.

Morphia was administered hypodermically during one summer The wound healed, leaving him a confirmed opium eater When he came under treatment, he was in the tenth year of the habit, and was consuming eighteen grains of morphia per day, taking it hypodermically. He was weak and emaciated, suffering symptoms usual to the habit. We had previously discovered *that patients consuming a given amount of morphia subcutaneously, can be sustained on the same quantity when administered through the stomach*, and avoid the excruciating pain caused by the pertusions.

Such persons often declare their inability to consume the drug by the stomach, without exciting vomiting We have to meet the first case of that character, which we could not overcome, as a large proportion of the persons who say this are humoring a notion. That many persons cannot tolerate or retain opium in the stomach, is a fact familiar to all physicians. It is owing to the presence of that peculiarity, that many are kept from

forming the deplorable habit. We are, however, writing of the drug as exhibited. The patient now under
consideration was of a lymphatic temperament, his secretions being readily acted upon, a fact which kept his
case free from complications other than the condition of
his old wound during the crisis, for which we entertained fears. Stopping the hypodermic injections, we
ordered for him the following mixture, No. 1 :

R.   Morphia, ℨ xii.
            Alcohol, ℥ v.
            Glycerine, ℥ vi.
            Gum Arabic, ℨ vii.
               Aqua, ℥ xxxv.
M. Sig.   Two teaspoonsful after meals.

Bottle No. 2 :

R.   Cannabis Indica, ℨ v.
            Alcohol, ℥ x.
            Ginger Tr ℥ x.
         Gentian comp. Tr ℥ xii.
            Glycerine, ℥ v.
M. Sig.   Replace what is taken from No. 1 with No. 2.

In stopping the hypodermic injections and taking prescription No. 1, he experienced no inconvenience. During the two years previous, he had taken wine and
brandy to ward off symptoms of vertigo and weakness.
We requested him to stop their further use, insisting
that our instructions should be rigidly enforced, as the

treatment suggested sustains the system when liquor and opium have both been used. He followed the instructions and improved rapidly. On the twentieth day he complained of disagreeable sensations, occurring early in the morning and during court sessions, especially when he was physically exhausted, his system seeming to demand the opium. We explained to him, as we are particular to do in all cases, that the symptoms were transitory, induced by changes taking place in the circulation as relaxation approaches, and warned him not to succumb to them, as the manifestation is abortive and erratic and not a genuine prostration, demanding for its removal sustaining remedies. By waiting, the system thrown momentarily from its equilibrium, will right itself. A drink of water, or a little food, facilitates its return to a normal condition.

If the patient honors this supposed demand, and takes his medicine before the appointed time, or indulges in a stimulating drink, the treatment is broken and a desire is created for a repetition of the indulgence, and thus the first determined resolution is destroyed. This minuteness of method may seem uncalled for, but the physician is fighting a desperate drug, involving the patient's after-life and the welfare of his offspring.

We ordered the patient to stop reducing his preparation No. 1 with No. 2 for seven days. The physician should always be explicit in giving instructions, and should adhere strictly to system. The rest improved the patient greatly. After again reducing with No. 2 for five days, he complained of prostrating symptoms ; but they were immediately followed by sensations of renewed strength.   We considered to give any but moral support to such symptoms, would be rendering that officious assistance, which is mischievous ; as the patient's system was assuming a normal condition.   Excrementitious waste was restricted, the system was appropriating nutritive principles and transforming them into adipose tissue. As these changes are in progress, and organs which have been obstructed by more or less plastic exudation are throwing off such hinderances, the nervous system, for a limited time, will reflex an action which depresses, its energies being concentrated in one direction ; but the general system will soon receive its proper nutrition. If not interfered with, nature will fill her storehouse with molecular deposits for future use.   We gave our patient frequent rests by stopping the use of preparation No. 2, and carried him to the forty-eighth day before the crisis appeared.   Relaxation of the bowels, during the last three weeks, indicated a favorable termination of treatment.

The crisis was ushered in by diarrhœa, vomiting anorexia and general prostration ; a large warm water enema was given to wash out the lower bowels, followed by a suppository containing ·

℞.   Musk, gr v.

Cannabis Indica, gr. ii.

Lactucarium, gr. v.

M.   We doubt the intrinsic efficiency of lactucarium

as a soporific, but it produces some effect in combination. With the aid of baths every hour, and mustard over the epigastrum, the patient was able to retain cicuta gr. $\frac{1}{18}$, musk gr. v. ; but we did not consider it prudent to administer phosphoric acid, because the appearance of the mucous coats of the mouth and fauces indicated susceptibility to irritating remedies, and consequent danger of gastritis. We had excised the uvula when treatment commenced as it was easily relaxed, exciting a disagreeable cough, and we knew that it would still further elongate during the crisis. Under the influence of musk, cicuta, nitre paper and baths, the patient rested quietly without the acid. The following day coca leaves were masticated and the saliva swallowed. He complained of darting pains in the old wound, settling down to a steady ache, similar to the pain suffered when the wound was active, and for which the opium had been taken, to allay wandering pains, lancinating in character, vacillating between the head, back and scapula. Ladies are apt to suffer terrible backache at this period, owing to uterine congestion. We ordered a liniment applied to vesicate the cuticle over the parts, consisting of :

R.   Oil turpentine.

    Acetic acid.

  Liniment of camphor ᾱᾱ $\mathfrak{Z}$ ii.

(Officinal). M. Sig. Rub the parts with the liniment thoroughly.

The baths were wonderfully soothing to this patient,

exciting sleep at once. We ordered that his body should be well greased, to avoid over absorption, and allowed him to remain, for an hour at a time, with a pillow under his head. He enjoyed perfect ease, and no bad effects were noticeable, his pulse being full and regular. Although in complicated cases patients suffer intensely, yet the pulse will remain normal. After allowing the baths to be persevered in, to the limit of prudence, a wet sheet was substituted, the patient responding to its quieting effects. The bran pack in such cases will also give good results, inducing sleep for two or three hours at a time.

The bath water should be changed often, as an immense amount of effete matter is thrown off by the cutaneous secretions, an offensive odor arises from the water used by patients who have taken opium for years. Our patient advanced finely, returning to the active practice of his profession the third day after his crisis. We had him continue the coca leaves, and also ordered quinia to be taken, one grain every three hours, not especially for its tonic action, although such an effect is secured, or, because it improves the tone of the stomach, by regulating peristaltic action, but principally for the reason that it protects the system against the results of certain tendencies, to which the patient is subject through an inability to exercise a normal power of resistance.

Quinia as a prophylactic, stands pre-eminently high. A noticeable phenomenon, observed when the system is responding to our methods, and the opium cachexia is progressively disappearing, impresses the fact upon us, that opium suppresses the functions of animal life, while apparently sustaining them. On the withdrawal of the last dose, patients suffer extreme prostration, and any one unfamiliar with their peculiar state, would give a dis mal prognosis; but on the following day they will sur. prise friends and physician by returning to business.

Five months subsequent to treatment, our patient presented no signs or symptoms, indicating that he had ever been a victim of the opium habit. His old wound gave him pain, however, for some weeks after the crisis.

CHAPTER VII.

CASE No. 5. Miss Lydia A., twenty-eight years of age, the daughter of a country physician, had opium administered to her for dysmenoorrhœa, and being familiar with the drug, and its dose, she continued its use on her own responsibility, until a habit was formed.

She had the appearance of being at least fifty years old, as no surplus tissue covered her bones. This, however, did not prevent the cadaverous look in the face peculiar to opium eaters. She could not carry on a conversation intelligently, but presented the foolish, vacant stare observable in epilepsy. Her history is peculiarly interesting and instructive, as it demonstrates some remarkable facts, with respect to the action of the drug.

At the age of fourteen years she had free access to her father's office and medicines, and understood therapeutics to a limited extent. During her first catamenial period, which occurred at that time, ten drops of laudanum were administered by her father, to allay the pain. Having a finely organized nervous system, she was fascinated by the exhilarating effects of the drug, and subsequently helped herself until a habit was formed. Her father died, leaving her a confirmed opium eater. Through her failing health and queer actions, her family discovered the condition she was in, and used every means known to them to break the habit, with the usual results.

Her mind soon displayed evidence of dementia, and her education was necessarily suspended. She was changed from a bright, vivacious girl, remarkably quick and aggressive, to one dull and sleeply, prematurely old and broken. Her friends removed her from school, as her intellectual faculties and capabilities seemed to be dormant, and during the next fourteen years, slow intellectual decay was observed. At the time when she came under our care she experienced difficulty in retaining even the knowledge already gained, while her receptive faculties were almost totally lost. She could read and write indeed, with the same facility as before she formed the habit, and execute pieces of music learned in girlhood, but the brain could retain no new impressions. Her case presented the remarkable phenomenon of a brain remaining stationary.

The case has no parallel in our experience with the drug. While the anomalous conditions created by its use are many, this one is not within the range of ordinary experience, and, as a rare exception, it is worthless as a precedent, it cannot be placed in the category of possible results of the habit in adults. While the brain is always progressively encroached upon, the degree of waste is self-limited.

This woman had increased her dose rapidly, until one-half drachm of morphia, and one-half pint of laudanum were consumed each day. She was in the habit of buying at wholesale, forty dollars worth of opium, which could be made to last her for one month. With that amount she suffered and pleaded for more, and she had taken this prodigious quantity for seven years. Her first menstruation proved also to be her last, constipation, with all the usual symptoms, being present also.

In putting her under treatment, we were confident that such a quantity of the drug could not be taken up by the system, and made to reflex its stimulating effects. The system laboring under unnaturally large doses, must necessarily suffer its acute depressing influence; we therefore compounded a bottle to last one day containing:

R.   Morphia, Ð i.
        Cannabis Indica, gr. ii.
           Alcohol, ʒ v.
             Aqua, ʒ i.

M. Sig. Three teaspoonsful morning and night, and two at noon.

Instructing the patient's sister to refill the prescription that evening for the following day, providing it proved adequate to sustain her system, and to report the result. To our surprise, no reaction was experienced, and the patient declared herself "cured," the relief afforded by reducing her enormous dose must have been decided, as that amount of opium, inflicted upon her system each day, could excite only sensations of pain, and discomfort.

Our experience with this case is associated with feelings of regret. We should have cured the woman, and our failure to do so occurred through a careless misapprehension with respect to nature's ability to repair her depleted forces, when undisturbed by contravening influences. A want of confidence in our own ability prevented, what would otherwise have been a brilliant victory. Her weakened state, both of mind and of body should have been considered, and the imperative necessity of not destroying her first intention by forcing her system beyond its powers of adaptation, should have been recognized.

We ordered preparations Nos. 1 and 2, compounded on a basis of twenty grains of morphia per day, for sixty days. Nature, relieved of its burdens, accomplished wonders for the patient during the first sixty days. Her mind seemed to be completely restored, enabling her to entertain company, and to shop intelligently. Her face assumed a natural appearance, and her weight increased fourteen pounds during the time. If her system had been allowed to rest every few days, favorable results would have been obtained beyond a doubt. But reaction was invited by our course, for on the sixty-fifth day, when the morphia was reduced to less than two grains per day, we found the patient had been taking laudanum for two days, obtaining it from her druggist, buying, however, but one-half ounce each day. That quantity was sufficient to destroy all moral influence over her. We should have watched for symptoms indicating the approach of relaxation, and should have suspended the use of preparation No. 2, until the system adapted itself to the reduction, resuming it after a rest, and proceeding slowly towards a cure. The patient's medicine should have been compounded in one ounce bottles, and numbered from one to ninety, respectively, allowing her to consume the contents of one bottle each day, and the cannabis indica should have been reserved for a later period of treatment, when it would have had greater effect. To secure accuracy in compounding the every-day bottles, Nos. 1 and 2 must be mixed in large bottles, and measured out into the smaller ones. Frequent rests for the system must be provided by suspending the use of No. 2. If the will-power of the patient is doubted, it is well to compound as above, putting the preparations in charge of a responsible person who will give them out as needed. By so doing, it is easy to avoid the danger of having instructions violated when passing sensations of discomfort create a craving for larger doses than are proper.

It is our custom to impress the necessity upon patients of confining themselves to the exact dose prescribed, any more or less being inadequate to allay nervous irritability. We direct patients also, if the dose does not have the desired effect, to report the fact, so that means may be used to relieve them. If their symptoms are indicative of undue prostration, the physician should allow them to stop No. 2, or should add more cannabis indica to No. 1. If the complaint is purely notional, it is well to administer a pill containing pulvis liquorice or capsicum, to satisfy the imagination, and to give patients who have a weakness in that direction, the satisfaction of knowing they are taking something. Had we taken this course the result would have been different in the case now under consideration.

Circumstances have rendered it necessary to place our patient under the care of a physician in a neighboring city, who is familiar with our methods of treatment. We fear negative instead of positive good will be the final result, through no fault of the present attendant, but through our own error in letting her miss the golden opportunity of a first attempt. When persons under treatment return to opium their maximum dose is soon reached.

CHAPTER VIII.

Case No. 6. Mrs. E. F. G., forty-two years old, while endeavoring to control the pain of neuralgia, resorted to opium, and thereby formed a habit. She was of sanguine temperament, and had a large osseous frame, with tendencies towards obesity. Previous to forming the habit she married at twenty-one years of age, giving birth to two healthy children during the subsequent ten years. While in gestation with her third child, neuralgia of a spasmodic character excited intense suffering; opium was taken to procure sleep, and the dose rapidly increased, so that she was consuming, when labor set in, ten grains of morphia per day. Her friends noticed on the third day after birth, that the babe was not receiving a sufficiency of breast-milk to sustain life, and the discovery excited surprise, this being so contrary to her usual habit, as she not only supplied an abundance of milk at the two preceding births, but also had galactirrhœa. The babe, bright and natural at birth, when placed on the bottle, became prostrate, vomiting and crying incessantly, discharging a glary mucous with the meconium from the bowels. The muscles twitched continously, and the babe died six days later, from an inability to take nourishment, and loss of sheep. The mother increased the dose of morphia gradually, until the birth of a girl babe, two years later, when she was consuming one-half drachm of morphia per day. The lactic secretions were entirely suppressed. The babe being bright and vigorous, took the bottle readily, but in two or three hours the mother observed the same symptoms that preceded the death of her last child, and called the doctor's attention to them. As it was bathed in a cold, clammy perspiration, and crying feebly and low, the physician was naturally surprised at so sudden a change, but could suggest no cause.

The symptoms becoming aggravated, the mother in an agony of fear comprehended, as by an inspiration, the true condition of her babe, the cause and effect presenting themselves to her. The babe had imbibed her blood, impregnated as it was by morphia. Its fibre having been held up, and contracted with her own, was now relaxed for want of its accustomed stimulant. The mother was called upon to decide a question, whose momentous results she alone fully understood. The life and the future of her child were at stake; she must act quickly and without advice or consultation, as her babe was dying.

She dared not confide her condition or fears to her husband, of whose real character she knew but little, although she had slept by his side for years; and fearing that her physician could not realize the exact state of the case, she did not feel warranted in trusting her secret to him. Years afterward she said: "Few mothers have passed through the mental agony I experienced; my maternal pains were a joy compared with it, not being over-shadowed by a terrible secret, for which I alone bore the blame." She made up her mind to purchase her babe's life at the expense of its being an opium eater, and sent the nurse for laudanum, then secretly gave it two drops. The effect was magical. The low, plaintive cry ceased, the cold, clammy perspiration that stood like beads on the forehead disappeared, spasmodic jerking gave way to natural sleep, and on awaking the babe was bright, taking the bottle again, for the second time since its birth.

The doctor was astounded to observe the change, and did not attempt to account for it. The mother gradually increased the child's dose, and it throve well, being remarkably free from sickness. The diseases peculiar to infancy were successfully passed without complications The laudanum was administered three times a day. When the girl came under our observation she was seven years old, and was consuming one-half ounce of laudanum per day. She seemed intuitively to feel that secrecy regarding her taking medicine was necessary, and never referred to the fact, but went to her mother at the appointed hours. If her father was present she would wait for an opportunity to take her dose unobserved. She was a bright, pretty child, and presented but few outward, or noticeable, signs or symptoms indicating that such a habit existed. Her ability to acquire learning from books was completely suppressed. She had attended school for three years, but was not familiar with the alphabet.

In placing her under treatment, we took into consideration the fact that she had always taken the drug, so that
its use was as it were a second nature to her. The blood
that sustained her embryonic life, had been impregnated
with it. A complete and radical change must now be
undergone by her system, as her present state was a
physiological one, having existed from the first molecular
deposit of her being, to the present time.

We relieved the mother, before treating the child. She
more thoroughly appreciating the nature and effects of
treatment, was an efficient nurse, familiar with the
necessities of the case.

The child's appetite was naturally good, requiring no
tonic remedies to excite those functions. We ordered
bottle No. 1, to contain:

R.   Morphia,   3 iv.
             Alcohol,   ℥ x.
                 Aqua,   ℥ xxx.

M. Sig. Two teaspoonsful after meals.

The amount taken was to be replaced every third day
from bottle No. 2, containing.

R Belladonna Tr. ℥ x.
                  Alcohol ℥ ix.
                Ginger Tr. ℥ x.
            Cannabis Indica, 3 iv.
              Gum arabic, 3 vi.
                  Aqua, ℥ xx.

M. Sig. Replace No. 1 every third day.

The child remained at school as usual, her parents being instructed to notify us when symptoms of relaxation made their appearance, but fortunately the advice proved to be unnecessary, as no crisis was experienced. On the seventieth day of treatment, prescription No. 1 was stopped and a tea of cocoa leaves substituted. The girl was restless for two nights, responding, however, to the soothing effects of warm salt water baths. Her bowels showing a disposition to remain relaxed, the extract of rubus villosus was administered, combined with a carefully selected diet, which succeeded in arresting the over activity. Her physical improvement was marked and rapid, not approaching, however, the effects her new condition wrought in the brain. She advanced in her studies with wonderful alacrity. Being ignorant of the pupil's habit, her teacher reported the progress made as phenomenal. This case illustrates the adaptability of the system in the young, and its capacity to accustom itself to decided changes systematically induced.

CHAPTER IX.

CASE No. 7. Mrs. Mary L., forty-two years old, had been married for fifteen years, having no children. She was of a peculiarly sensitive and highly nervous temperament, predisposed to sudden convulsions of feeling, with hysterical mania, which plunged her into fits of melancholy or exaltation. She was frail and weak, and a religious monomaniac. Such was reputed her condition previous to forming the habit.

Inheriting wealth from her father, she became, for the want of control, stubborn and set in her many erratic notions. She was highly educated, however, and a brilliant conversationalist. While we were familiarizing ourselves with her case, and endeavoring to win her confidence, she favored us with a specimen of her abilities in that respect of quite a suggestive character, delivering a lecture impromptu on her pet hobby, abounding in brblical lore, interspersed with quotations which required no reference. At the end of it she fell back exhausted, after an hour's incessant talking.

It is unfortunate indeed, for a person naturally inclined towards this state, to acquire the opium habit, as even well-balanced minds have to struggle in avoiding the hysterical tendencies superinduced by opium. Her spine was supposed to have been injured by a fall from a horse, but no positive evidence declared such to be the fact. Nevertheless, she insisted that a bone had been broken, rendering it impossible for her ever to walk again. Her friends humored this whim by letting her remain in bed. Morphia was administered hypodermically, and a habit quickly formed. She was consuming twenty grains of morphia per day, having it inserted one grain at a time every half hour during the day. No one. could use the hypodermic syringe to her satisfaction except her husband, a fact which rendered it necessary for him to give up his whole time to attend her bedside and avoid exciting her She was reduced to a skeleton-like appearance, retaining, however, extraordinary nervous energies and mental vigor. The special senses were morbidly acute and sensitive to outward impressions, creating a strong belief with her ignorant and superstitious neighbors that she possessed supernatural powers. She was accustomed to predict the approach of visitors before others were aware of their coming. She would take a sealed letter and do some very clever guessing as to its contents. We account for such freaks by accrediting them to the abnormal distribution of nerve force, which is apt to mislead the over-confiding who accept as supernatural phenomena incomprehensible to them. Physical activity being suspended, there naturally follows a concentration of nerve force, causing a sort of hypernutrition of some one of the special senses.

In this case it stimulated the perceptive faculties to a

degree not possible in health, enabling her to astonish her friends. Her arms and hips were in a terrible condition, owing to the great number of punctures made by the hypodermic syringe. The surface was one mass of indurated nodules, phlegmonoid swellings, and indolent abscesses. The muscles appeared to be atrophying, and, hyperæsthesia being present, she suffered intense pain. The smallness of the amount of food she consumed was the wonder of her friends. Having no appetite, she seemed to have a pride in eating so little and attracting the surprised attention of others.

In deciding upon our plan of procedure, we determined that an aggressive method of treatment must necessarily be preceded by a moral control over the patient's emotional tendencies. Her resolution must be sustained by full and implicit confidence in our ability. We felt impelled to make a desperate effort in her behalf, as her peculiar state involved the happiness of a noble, self-sacrificing husband. By being cautious not to antagonize her, especially when dealing with non-essentials, we succeeded in winning her confidence far beyond our most sanguine expectations. She could not retain medicine of any kind in her stomach, and fearing that there was irritability, we would not risk our first effort being a failure by adopting uncertain methods. We began active treatment in the morning by washing out the lower bowels, and inserting a suppository containing :

R̊.   Morphia, gr. vii
             Belladonna, gr. ⅛.

M.   This was to be passed as far up as possible, and followed by a sponge bath of warm salt water.   The arms and hips were bathed in :

> ℞.   Carbolic acid, ℥ i.
>   Glycerine, ℥ i.
>   Rose Aqua, ℥ ii.

M.   She was surprised and delighted to find that she could go without the hypodermic administration of morphia.   The suppository was repeated that evening, containing:

> ℞.   Morphia, gr. v.
>   Gentian, gr. ii.

The druggist was instructed to reduce the morphia at the rate of one-eighth of a grain per day, and to add cannabis indica, of the same weight as the morphia that was deducted.   After the fifth day her improvement was strikingly apparent to all, the suppositorys sustaining her perfectly.   Her appetite bordered on the ravenous, and the gastric functions assimilated the aliment in a normal manner.

The necessity for getting her out of bed, and having her take medicine by the stomach, was apparent.   A short ride being proposed, she strenuously objected, on the ground that she was unable to walk a step.   We insisted, and after reminding her of what had been accomplished, she consented to be carried out.   Before starting, the following prescriptions, Nos. 1 and 2, were compounded, enabling her husband to have No. 1 with him:

℞.   Morphia, ℨ x.
            Glycerine, ℥ vi.
            Alcohol, ℥ v.
            Aqua, ℥ xx.

M. Sig. Two teaspoonsful after meals.

According to a prearranged plan, the patient's first ride embraced the mid-day hour, and the time for a fresh suppository, at one o'clock. When some miles from home, she was informed of the necessity of having her usual stimulant, and of the impossibility of inserting a suppository, and was told that she must make up her mind to retain No. 1 taken by the stomach. She did so without exciting disagreeable symptoms of any kind, demonstrating that the nerves' periphery presiding over the gastric functions, had resumed its normal action.

Daily rides were enjoyed, and marked benefits received from them. The attendants were instructed, when assisting her out of the carriage, to allow her to stand unsupported for a moment, encouraging an effort to stand alone. After accomplishing that feat to our mutual satisfaction, a railing was temporarily erected, leading from the stile blocks to the door, enabling her to walk from one to the other, supporting the weight of her body by resting her hands on either railing. She was induced to make the effort, and surprised every one by going up and down-stairs at will, within ten days. Each succeeding day was marked by rapid changes for the better, the patient showing fewer signs of hysterical tendencies, and becoming more reasonable and natural.

We considered that the time had come when a candid exposition of her case would be fully appreciated, and have a salutary effect. We enlarged upon the dangers if a failure occurred, and the benefit following a successful treatment, and this had, as expected, a beneficial in·fluence. She recognized the precarious state the opium had placed her in, rightly estimating its dangers, and the imperative necessity of adhering to instructions. On the thirtieth day of taking No. 1 by the stomach, morphia ℥ i., with cannabis indica ℥ iii., were added to No. 1, with instructions to replace No. 1 with No. 2 every third day, as her subsequent condition indicated the critical period had been successfully passed, allowing her to discontinue all medicines on the ninetieth day, without exciting complications of any kind.

With nearly all hysterical cases mustard baths and pediluviæ give better results than salt water, and they were resorted to with this patient. When similar cases apply for treatment, presenting few encouraging symptoms, with a dismal prognosis suggested by their aggravated condition, duty demands that the physician shall make a well directed effort to accomplish their cure. He will find it necessary, in many cases, adroitly to utilize his diplomatic powers in overcoming the patient's caprices, without appearing to be aware of their presence. Something more is required of the physician, if he is to be successful, than a profound knowledge of the subject, as confidence must be won to give a permanent foundation to build upon.

The opium in its action upon this case added one more to the long list of its anomalous effects on the system. As a young lady our patient had labored under an inherited dyscrasy, producing complications that generally remain, according to the ordinary experience of the profession, as insurmountable barriers to good, or even a fair degree of health, consigning the sufferer to a life of abnormal nervous activity, and the consequent exhaustion of those functions that preside over nutrition by a continual expenditure of force in sustaining excitability. The opium not only intensified diathetic tendencies, but developed complications peculiar to itself. After eliminating the drug from the system, the forces controlling animal life assumed a degree of activity which might well have been considered impossible with her constitutional predisposition. The vital forces seemed to have imbibed renewed life, by being so long imprisoned. Her bowels became regular, when previous to taking opium, three weeks would pass without an action. Layers of adipose tissue were accumulated for the first time, and five months after her cure she presented the appearance of a healthy, robust woman.

CHAPTER X.

CASE No. 8, Mr. G. C. D., lawyer by profession; thirty-eight years old. His general health had been good until his thirty-first year, when sciatica followed a severe cold contracted in the spring, while hunting. Opium was administered to control pain, and a habit was formed. As usual the drug checked the intensity of his pain for a limited time, but ceased to have the desired effect after the habit was formed. During the third year of his disease he went to the Hot Springs of Arkansas, receiving complete relief from sciatica. He made a desperate effort to stop the use of opium, but to no purpose. He expressed his horror of the habit by declaring that sciatica caused him acute suffering, but that it could not approach the lingering agony of the opium habit. A remarkable feature of this case was that the patient had indulged the habit for three years, and yet was consuming but three grains of morphia per day. Nevertheless he was suffering all the symptoms usually resulting from excessive doses, his face showing the opium cachexia. We confidently believed that he was laboring under a mistake regarding the amount of his dose, as he went to sleep while describing his case. His account, however, corresponded with a letter received from his family physician, leaving no room for doubt. We were misled by the amount consumed, and did not exercise our usual caution, considering him susceptible of a rapid cure. We should not have accepted any condition connected with the opium habit as an excuse for relaxing our con stant vigilance.

Feeling responsible for the subsequent misery sustained by our patient and his family, we hope our mistake will remain as a warning to others never to take anything for granted, which may be reduced to a matter of certainty, for the vagaries of nature in this disease are many. It is best always to proceed with that studied caution that should always be inspired by the consciousness that on the physician's judgment a human life may depend. Our patient revealed an important fact in connection with the opium habit in giving a history of his case. He said he had drank liquor to excess during his early life, but had stopped the indulgence before forming the opium habit, not drinking again until a supposed demand for alcohol had arisen during the last two years. When his system was prostrated by the opium, four or five drinks of liquor per day sustained him, allaying symptoms of weakness and vertigo. We instructed him to stop at once the liquor and the opium, assuring him he would not desire either unless he wilfully indulged, and we prescribed the following:

R̠   Morphia, ℨ i.

               Glycerine, ℥ v.

            Belladonna, ℥ iv.

                 Aqua, ℥ viii.

M. Sig. Three teaspoonsful morning and night, and two at noon.

Bottle No. 2 contained:

R.   Cannabis Indica, 3 v.
            Alcohol, ℥ vi.
            Ginger Tr. ℥ iv.
        Gentian Comp. Tr. ℥ v.

M. Sig. Replace No. 1, as consumed.

His system should have been rested by stopping No 2 every few days, instead of constantly reducing the strength of the dose at the risk of shock to the nervous system. Withal the patient did well, gaining flesh at the rate of three pounds per week, and feeling no desire for liquor or opium. The bowels relaxed on the twentieth day of treatment, and the patient went into the crisis on the twenty-ninth day, with all the symptoms indicat·ing a speedy cure. He made few complaints, yet baths, phosphoric acid, musk, coca leaves, and nitre paper were used in the order heretofore suggested. He passed the first day and night in comparative comfort, being nearly out of the crisis by four o'clock the following morning, when he invited disaster by ordering the atten-dant to bring up a "whiskey sling," and as he was stay··ing in a large hotel he got what he called for.

At nine o'clock when we called, he was contented and free from nervousness. We observed, however, at once, that he had been drinking, and went into full details regarding the result which must follow, if he did not stop at once. This he promised faithfully to do. The following night a professional nurse assumed charge, with positive orders that no liquor should be taken. The night was passed in quiet until three o'clock, when the demon that he had awakened the previous night began to assert his supremacy, and our patient demanded liquor. The nurse implored him to resist the desire until we could be sent for, but in vain. Being his own master, he received what he paid for, and took four drinks of liquor. He then wrote a letter, thanking us for our kindness, and closing by assuring us that he would not drink after arriving at home, and started on an early train.

His family physician wrote a congratulatory note, declaring that "our patient was looking well, but had the appearance of drinking" We answered immediately, giving a full history of his case, and expressing our fears for the future, advising the doctor to put him on infinitesimal doses of opium, and to go over the ground again ; but the patient would not listen to the proposition, vehemently declaring himself "cured," and insisting "that liquor would not get the start of him."

Eight months afterward we received a letter from him, the orthography of which clearly indicated his condition. He had been going to fearful excesses in liquor, and begged for help. We advised that he should place himself under the care of his family physician, but he arrived three days later unannounced. He was bloated, presenting all the extrinsic symptoms of confirmed dipsomania. We considered it a useless waste of time to treat him for that complaint, as the same nervous conditions were not present that are created by excesses in alcohol, independent of opium. We gave him a preparatory treatment, however, not wishing to confine in his system the accumulation of effete matters resulting from excesses in alcohol, by suddenly closing his secretions with opium. We administered calomel of the tenth trituration, as follows :

℞.   Calomel, gr. x.
           Podophyllin, gr. $\frac{1}{11}$.
           Capsicum, gr. i.
           Gamboga, gr. i.
M.   Pill i.

Three hours afterward a large enema of soap water was given to empty the lower bowels, and this was followed by a warm bran-pack. Fluid extract of Juborandi was given in teaspoonful doses every half-hour, until a profuse diaphoresis was excited. The following morning, after obtaining the desired effect from the medicines administered, we put the patient on Nos. 1 and 2, No. 1 containing :

℞.   Morphia, gr. x.
           Glycerine, ℥ v.
           Alcohol, ℥ ii.
           Aqua, ℥ x.
M. Sig.   Three teaspoonsful after meals.
No. 2 containing :

℞.   Cinchona Rubra, ℥ x.
              Ginger Tr. ℥ v.
        Gentian comp. Tr. ℥ vi.
            Capsicum, ℨ v.
M. Sig.   Replace No. 1 as before.

He had no desire for liquor after taking the medicine, and went safely through the crisis on the twentieth day. Profiting by experience, he subsequently let stimulants of that class alone.

CHAPTER XI.

CASE No. 9. *Suggesting the treatment of a pregnant woman necessary to save both mother and child.* Mrs. A. M., twenty-two years old, was married at sixteen. A painful uterine trouble was developed after marriage. In going from physician to physician to obtain relief she formed the opium habit. Her disease was finally cured by a celebrated New York specialist, performing an operation on the cervix. A year afterward she discovered herself to be with child, and was also a confirmed opium eater, in the third year of her habit, consuming twelve grains of morphia per day. The effects resulting from its use were well marked. She was very anxious to be cured, not only on her own account, but also on that of her babe to come.

The indications were to reduce her dose to the smallest amount possible—consistent with a quiescent state of the uterine functions—and to avoid relaxation of the system. That would excite expulsive effort and produce abortion by a reflex action under the influence of a reduced dose. The young woman soon to be become a mother is called upon by nature to perform a double function, and give life to the fœtus as well as sustain her own being. Nature must be unchecked if it is successfully to carry out the phenomena of reproduction in the brief period of nine months, and, by the elaboration of blood through the mother, to sustain embryonic nutrition.

Reducing the dose of opium with such patients sets in motion the primary functions for accomplishing the extraordinary achievement. The action and influence exerted by opium over the menstrual functions are rich in physiological suggestions. The woman consuming twenty grains of morphia per day will menstruate once in eight months on an average. Yet she is susceptible of impregnation with the same degree of certainty as are women free from the habit, a fact which demonstrates an interesting truth. No catamenial fluid is discharged, yet the ovules, or germ cells, present themselves with unvarying regularity, proving by their ability to become fecundated that it is a normal manifestation.

The necessity of rapidly reducing the opium taken during the early months of gestation, is very great, not only for the purpose of improving the general nutrition of the mother, by increasing and enriching her red blood corpuscles, but also to prepare her for sustaining her babe by supplying a sufficiency of milk, containing the requisite constituents to maintain molecular activity. Great care must be exercised in reducing the opium under such circumstances, the physician taking ample time, and making no attempt to withdraw the last grain until after the child is born.

While nursing, the last grain can be cautiously withdrawn, curing mother and babe at the same time. If the usual dose is not systematically withdrawn by appropriate treatment, the lactic secretions will be scant, or wanting, such a state will materially affect the after life of the child, because of defective nutrition, which instigates complications difficult to allay by reason of their obscure character. We must give the patient the benefit of tonic remedies as the dose of opium is reduced. Under favorable surroundings the improvement will be so encouraging as to mislead the physician regarding the amount necessary to sustain the patient until after labor.

One should not hazard the attempt to cure such a patient by stopping the last grain, however confident may be the patient's hope of success, founded upon her apparently favorable condition. It is difficult to comprehend the effect of a grain of morphia on the system which is habituated to its use. If a premature attempt is made, the clear sky is soon overcast with clouds, whose blackness and swiftness of approach strike terror to the heart. The patient, so happy and confident, anticipating no trouble, supported and sustained by her confidence in the physician's skill, suffers relaxation, excited by going too fast in the attempt to relieve her entirely. The uterus expels its contents in a remarkably short space of time, the flood-gates of the system are opened if the physician is not watchful and expeditious in getting control of her nervous system. A hemorrhage from the relaxed uterine plexus will, by its terrible severity, carry the patient off before there is time in which to comprehend the situation. It is imperative, therefore, that the physician should not let a serene exterior mislead him.

We placed case No. 9 under treatment by prescribing Nos. 1 and 2.

As the woman was in the fifth month of gestation, we compounded No. 1 on a basis of twelve grains of morphia per day, for sixty days. No. 1 contained :

R.　　　　　Morphia, ℥ xii.
　　　　　　Glycerine, ℥ x.
　　　　　　Alcohol, ℥ x.
　　　　　　Aqua, ℥ xx.

M Sig. Two teaspoonsful after meals.

No. 2 contained .

R.　　　　　Cannabis Indica, ℥ vi.
　　　　　　Quinia, ℥ iv.
　　　　　　Alcohol,
　　　Gentian comp. Tr. ℥ x.
　　　　　Ginger Tr. ℥ vii.
　　　　　Glycerine,
　　　　　Aqua, aa ℥ vi.

M. Sig. Replace No. 1 every other day with this.

She improved so rapidly during the subsequent two months that a condition approximating health was attained. Bottle No. 1 was refilled by adding morphia, ℥ iv. No. 2 was also refilled with cannabis indica, ginger, and gentian as before. Going the full term, she gave birth to a well developed child, and supplied an abundance of milk to nourish her babe. We rested her system during labor, and the following two weeks, by stopping the use of No. 2, allowing her dose to remain uniform.   Resuming No. 2, and reducing No. 1 with it each day, we were able to withdraw all medicine twenty-eight days afterward without exciting a crisis.

As a precautionary measure, it is judicious not to enlighten women in this condition regarding their liability to abort, unless they evince some anxiety for the safety of the babe.   A strange inconsistency is displayed by many who show no love or solicitude for the life they carry and nourish, but are often willing to run a gauntlet of dangers in order to rid themselves of that which they will love and cherish after experiencing maternal pains, and to the well-being of which they will willingly devote life and happiness.

CHAPTER XII.

CASE No 10. *Treatment pursued with a babe twenty-six hours old.* The mother having been addicted to the opium habit for four years, went to great excesses during gestation, consuming three ounces laudanum per day. The secretion of milk being suppressed under the drug's action, the nursing-bottle was necessary. The babe, although frail, evinced no lack of vitality, and took the bottle readily during the first twelve hours, when symptoms appeared indicative of relaxation for the want of the accustomed supply of opium which the babe had heretofore received through the funis. The appearance of these symptoms corresponded in time with a similar condition sustained by adults going without their opium, twelve hours being the maximum duration of its action.

The babe became irritable, refusing the bottle, crying and moaning incessantly, and passing from the bowels a dark, mucous-like substance, with constant vomiting. On coming under our observation the symptoms were becoming greatly aggravated, and were combined with spasmodic nervous action affecting the muscles, also ataxic symptoms suggesting a speedy dissolution. The extremities were cold, and covered with a clammy perspiration. We attempted to administer ten drops pargeoric with three of ginger in a little milk and water, by passing it well into the mouth with a dropper, a large proportion being lost through spasmodic effort of the muscles presiding over deglutition. We gave three drops of laudanum by inunction over the stomach, enabling that organ to retain the paregoric—the system requiring a teaspoonful before a natural condition was secured, when the babe nursed its bottle and went to sleep.

It was important for us to ascertain what was the smallest amount of opium the babe could take and maintain life. By carefully watching our little patient, administering a few drops of paregoric at a time, increasing the quantity as symptoms demanded, we found that three teaspoonsful of paregoric would sustain it for twenty-four hours. This amount was given in three doses of a teaspoonful each, morning, noon, and night, in milk. No change was made during the first ten days, in order to allow the child time in which to gain strength. Then one drop was taken each day from its daily supply of three drachms, and we ordreed a rest of ten days to be taken after eighty drops had been withdrawn, the rest to be repeated when a drachm, or sixty drops, remained.

In withdrawing the last thirty drops the reduction of a drop was made only every other day. The babe thrived during its treatment, and no crisis was noticeable. The mother's care presented no points of special interest, she being fully relieved during the subsequent three months, without complications, the only abnormal symptom remaining being incontinuance of urine.

CHAPTER XIII.

CASE No 11 *Showing the relations existing between the opium habit, and pulmonary tuberculosis.* Mrs. D. K., twenty-two years old, married for six years, with one child, inherited a tuberculous diathesis from both parents. Her father died at twenty-five years of age. Her mother displayed well marked symptoms on physical examination of progressive structural changes, involving the lungs, incident to consumption. Our patient had phthisical symptoms early in life. She sustained a hemorrhage, and had the characteristic cough. After the birth of her only child, the symptoms denoting consumption became intensified; a continual hemoptysis drew heavily upon her impoverished resources, threatening an untimely termination of her case.

In testing the efficiency of different remedies, and their ability to control her cough, morphia was taken, and for a time received the credit of being the long-sought-for specific, as it not only soothed and quieted nervous symptoms, but allayed as if by magic, the alarming phthisical symptoms. The cough and profuse expectoration ceased, encouraging the belief that a cure had taken place.

Without divining the cause, she discovered that no more milk was secreted, and weaned her babe. For a few months, and before she had increased the use of opium to a point incompatible with good health, she gained in strength and weight. Could she have con-trolled the tendency to increase the dose, few debilitat-ing results would have followed. But she increased the dose rapidly, soon suffering all the terrible consequences of excess.

On coming under our observation, she was consuming thirty grains of morphia per day. We placed her under treatment, by prescribing Nos. 1 and 2, compounding No. 1 on a basis of fifteen grains of morphia per day We also directed a vigorous treatment to check the con-sumption. This step we considered a necessary con-dition of even tolerable success in relieving her of opium.

The word consumption is losing its power to appal the physician. In times past, a favorable geographical location was considered the great desideratum of treat-ment. A radical change has taken place. We now treat the disease, instead of wasting our efforts in mak-ing doleful prophecies.

The physician should disabuse his mind of all pre-conceived notions that even suggest that consumption can not be cured, and by calling into active requisition the faculties that God has given him, prepare to lay siege to the monster in his den. We make a few sugges-tions regarding the classification necessary to form a basis of intelligent treatment, pointing out character-istics and symptoms.

As a disease, consumption requires for its relief decided measures of treatment ; powerful naroctics are often indicated or desired by the patient, rendering this important topic in all its bearings, of paramount interest to us, as its therapeutics are closely allied to our subject.

The results attained by the following treatment were decidedly favorable, and unequivocal. The general practitioner has presented for his consideration three well defined conditions, each having distinct pathological lessons, pecnliar to itself, which should be recognized in order that a plan of treatment may be based upon the particular facts of the case.

Until within a few years, consumptives were given the stereotyped advice to go south, east, or west, and it was thought they were doomed in any case. Such a theory was even the foundation of teaching in our medical schools. The theory discourages the aggressive and intelligent methods of battling with a disease that should be characteristic of the physician.

We do not wish to create the impression that we oppose a change of climate ; we are strongly in favor of such changes, believing them to be conducive to the arrest of tissue waste, under certain conditions and circumstances. We are free, however, to say, that the manner of making the change adopted by a large majority of patients, is radically wrong, while the localities are chosen with a total disregard of all the essentials of purity, density, or rarity of the atmosphere. The effect is thus rendered pernicious in the extreme. Nineteen out of twenty persons who have gone away, would have done better if they had remained at home. Under the head of treatment, we will refer to favorable places, and the manner to be pursued in going.

One of the manifestations embraced under the head of consumption that is amenable to treatment, especially in its earlier stages, also in the form of fibroid phthisis, follows chronic bronchitis. A controlling and curative influence can also be exerted over the transmitted, or diathetic diseases, pulmonary tuberculosis, and catarrhal pneumonia.

There are certain principles of treatment, applicable to all cases, including measures to promote digestion and assimilation, to increase the force of the circulation, and to create favorable hygienic surroundings.

Pulmonary troubles are generally preceded by complications that weaken and lower the tone of the general system ; the patient's nutrition is not up to its general standard, the gastric functions imperfectly perform their part, and the sufferer begins to droop. Persons thus situated, fear dyspepsia ; the appetite is capricious, being unnaturally acute, or entirely deficient, the patient usually having a distaste for fatty articles of diet.

The processes of repair, whose normal action is to replace constant waste, are not equal to the task, and the organs charged with this function become progressively anæmic, creating the primary conditions for exciting pulmonary diseases. The circulation is lowered, the arteries have lost, in a measure commensurate with the ·amount of anæmia, their contractile powers to carry the blood to the peripheral vessels, with that force which is necessary to health.

When the system, for the want of nutrition, has arrived at a point involving impaired circulation, pulmonary complications are developed, whether the disease has declared itself or not. Measures of treatment must be directed, so as to excite assimilation and improve nutrition. If the patient displays an aversion to fatty substances, it is no time to prescribe cod liver oil, as the insulted stomach will rebel, forever destroying the ability to take that remedial agent.

The physician should endeavor to correct the morbid state that excites such unnatural tendencies. When the patient experiences a heavy feeling in the stomach after meals, and does not desire food, it is well to withdraw all aliment, except milk, instructing him to persist in its use, until a natural appetite returns, which will appear about the fifth or eighth day

The stomach, in the meantime, has taken a needed rest. For this condition, we have applied the actual cautery, down the spine, on either side of the vertebra. As the appetite returns, farinaceous articles of food, with those containing a large percentage of saccharine matter, together with fruits, including one or two lemons each day, may be taken with marked benefit.

If the tongue is coated, it is indicative of that low form of inflammation, observable in dyspepsia, calling for Tr. iodine, in three drop doses. three times a day, well diluted in water. We have administered oil-and brandy, in combination, after the following formula, without exciting nausea, even in patients who had an aversion for oil. In this way we avoid at the same time the danger of cultivating a taste for the liquor.

℞.   Brandy, ℥ xii.
Glycerine, ℥ iv.
Cod liver oil, ℥ vi.
Simple Syrup, ℥ ii.
Fluid Ext. Coffee, 3 v.
Gentian Comp. Tr. ℥ iv.
Peruvian Comp. Tr. ℥ iii.
Tr. Ginger, ℥ ii.

M. Sig.   Wine-glass full after meals.

The vomiting of phthisis denotes peripheral irritation of the nerves. Carbolic acid, compounded with glycerine, will often allay the symptom. We have enabled patients to retain a meal, by administering before the taking of food—

℞   Chloroform, ℥ ii.
Capsicum, 3 i.
Tr. Cubebs, ℥ i.

M. Sig.   One-half teaspoonful before meals.

Aliment, as represented by oils and brandy, including tonic remedies, should be invariably given after meals, or with the food, so as to be emulsified with it. Innutritious liquids, such as tea and coffee, should be interdicted, as they retard digestion, and arrest deposits of nutrition, that enrich and increase the quantity of blood, and improve the circulation.

It is necessary to remove, so far as possible, all ab-normal products accumulated in the lungs. This must be accomplished by direct treatment.

In caseous phthisis, infiltration has taken place, ac-cumulating in the alveole. To facilitate its removal, we have devised a plan for throwing powder into the bronchi, by fastening a flexible rubber tube to a glass one, putting the powder to be inhaled into the latter. Passing the rubber tube well down toward, and as near to the larynx as is possible, without exciting reflex ex-pulsive effort, we direct the patient to exhaust the air from the lungs, and then to take a deep inspiration. At the moment of inspiration we forcibly blow the powder through the tube, enabling the lungs to appropriate a large proportion of it. The powder to be first used consists of :

℞.   Argentia nitras, gr. ii.
              Pulvis Peruvia, ℈i.
M. Triturate for ten minutes.

Sig. Insert one-fourth of the above quantity, every morning and evening. After persevering in the powders for three or four days, a closed compartment can be arranged, by the aid of five rubber blankets leaving room below for the entrance of fresh air, the patient assuming a comfortable position within. The compart-ment is to be filled with steam, surcharged with carbolic

acid, the strength being one of acid to forty of water.
A rubber tube conducts the steam from a boiler to the
compartment, the patient remaining in this steam from
two to five hours each day, as the circumstances attending each individual case may demand.

When the physical signs indicate the presence of large
cavities, and a rapid waste of lung tissue, it will be necessary to charge the steam with chlorinated lime, $\frac{3}{}$ i. to
a pint of water, having the patient remain twenty
minutes in the steam at a time. If he experiences a
sense of pain in the lungs, after using the latter, he
should enhale the steam from the following preparation
for its soothing effects:

> ℞.   Borax, $\frac{3}{}$ v.
>   Glycerine, $\frac{3}{}$ ii.
>   Aqua Rose, quart i.
> M. For two or three hours.

We have also used the steam from Tr. iodine and
Rose water to good advantage. When the patient
presents on examination, cavernous respiration, denoting that a portion of the lung has solidified, with a cavity
in close proximity, inhaling the steam from the salts of
ammonia, will be indicated.

The degree of success ultimately attained will depend
greatly upon the accuracy of the diagnosis, and a proper
appreciation, on the physician's part of the extent and
character of the lesion presented.

The clinical history of each condition is sufficiently.

diagnostic to be discriminated, suggesting the strength and activity of treatment necessary, and the methods should be determined according to the circumstances of each case.

The cough will subside, and fever, if present, will abate. The cough is sympathetic, and is a local expression of a complication, not amenable to the action of cough syrups, and their use should not be recommended. Cutaneous activity should be stimulated by friction, and salt-water baths.

When no fever exists, exercise can be taken; when a tendency to fever is present, the patients should be carried into the open air, having them make as little effort as possible, as tissue waste is going on at that time, to a much greater degree than in any other condition. Night sweats in consumption must receive careful attention. It is highly important to check this constant waste, as organic matter, necessary to the nutrition of the body, is progressively lost, counteracting good results attained. Night sweats are occasioned by a relaxation, superinduced by a paralyzed state of the nerves presiding over cutaneous activity

We find good results follow a bath taken before retiring, drying the surface with a coarse towel, vigorously rubbing for ten to fifteen minutes. It is well also, to let a current of electricity pass through the water. Atropia combined with dover's powder will often control the sweats, while the usual effect of the latter is to excite perspiration. Picrotoxin has proved efficacious, in the hands of many persons. Hemorrhages, or the spitting of blood, can be controlled by smoking, or inhaling the fumes of opium, but we can only recommend the use of opium as a temporary expedient, while the physician is getting the upper hand of the disease, by methods that produce permanent results. Efforts to improve the circulation, applicable to most cases, consist in the use of tonics, friction baths and pure air. Cannabis indica has proved in our practice a tonic capable of increasing the red blood, and of improving the circulation, by giving tone to the vasomotor nerves. If pneumonia attacks persons predisposed to weak lungs, the opium in full doses, if tolerated in the form of dover's powders, can be relied upon. If a sympathetic fever is present, juborandi, assisted by bran-packs, will be indicated. During the stage of softening, the steam from Tr. iodine, with opium at night, can be persevered in. With patients suffering the last stages of pulmonary consumption, and suggesting by their appearance that a fatal termination is near at hand, transfusion should be boldly resorted to, and the vital energies restored, by injecting into the venous system, blood taken from another person. Milk, instead of blood, has, in the hands of an eminent English physician, given good results, injected in the same manner, into the middle basilic vein. We will not refer to the operation, as the profession have been familiar with its details since the sixteenth century. We should endeavor to discover wherein our predecessors have failed, and to profit by their experience. It is not consistent with the conservative work-ing of nature to assume, even under auspicious circum-stances, that the system can appropriate nutritive matter to the extent of four or five ounces in as many minutes. Yet operators sometimes inflict that amount upon it, and declare the method a failure, because the organs cannot assimilate the amount. We can see no necessity for drowning a person for the sake of giving him a drink. As before demonstrated, the system is susceptible of rapid changes, systematically induced, and when we imitate nature, in supporting her with nutrition, trans-fusion will be a success. The blood or milk should be slowly injected, in small quantities, allowing time for its appropriation; we must not paralyze the system so that vital reaction cannot take place. It is best to use both arms, following minutely the instructions given by our best and most recent authors.

The choice of a suitable climate for the consumptive, and the grave responsibility imposed upon the physician, in deciding a question so fraught with possibilities of good or evil, should be cautiously considered. The physician should take into careful consideration his patients' needs as exemplified by their condition, other-wise grave errors are liable to be made. The physician must thoroughly understand the condition of his patient, the atmospheric surroundings of the place recommended, and the proximate effect it will have upon the particu-lar case in hand. He should estimate the fatigue in-curred by a long journey, and discover whether the possible results to be gained will be equal to the case involved. The comforts of a quiet home life should not be carelessly sacrificed, for uncertain and remote benefits, supposed to be obtainable.

The condition calling for a change, is the necessity for a rarefied atmosphere, to accomplish what the physician has faithfully endeavored to do at home without success; to excite an increased and rapid circulation, to carry the life-sustaining fluid to the peripheral vessels, there to deposit molecular masses of nutritive matter, to supply tissue waste. If such results cannot be obtained at home, a change is to be advised, into the elevated regions surrounding Denver, or Pike's Peak, Col. One should not send patients there at a bound, however, and kill them at once, by a sudden alteration of their usual habits of life. The removal of atmospheric pressure that they have been accustomed to, is apt to superinduce a fatal hemorrhage. The physician should allow them to go a part of the way by rail, if they desire, but advise them to finish the journey by wagon, at the rate of fifteen miles per day. This gives the lungs an opportunity to adapt themselves to new surroundings, by securing pure air in abundance, and all the conditions essential to a restoration of health. We can not refer with confidence to other localities.

Patient No 11 improved rapidly, as the opium was decreased, gaining in flesh, presenting within twenty days quite a robust appearance, the phthisical symptoms yielding to the combined exhibition of argenti, nitras, and carbolic acid steam. Her case being an aggravated one, she was kept in the steaming compartment for six hours per day, for three consecutive days, and all unfavorable symptoms disappeared. We found by careful experiments, that three grains of morphia per day were necessary to ward off relaxation, and maintain an active state of all the patient's functions. Phthisical symptoms also appeared, when a reduction below that quantity was made. We therefore ordered her family to administer that amount regularly, in conjunction with cannabis indica, ginger and gentian, making no effort to reduce the dose for several months. We can give the reader a more perfect and intelligent idea of her condition, by recapitulating; enumerating the important points involved, and the results attained by treatment.

On presenting herself for treatment, the signs denoting tuberculosis, advanced to the second stage, were well marked. They included a depression at the summit of the chest, showing a diminished volume of lung at the apex. Percussin elicted flatness, also tympanitic, amphoric, and cracked metal resonance indicating the presence of cavities, and solidification, with the characteristic emaciation, cough, and expectoration. Her previous history showing that morphia had been taken at first in small doses, followed by marked improvement, the dose was rapidly increased, and within eight months twenty grains were consumed per day. Under the influence of increased amounts, serious complications, peculiar to the drug, were excited, prostrating the nervous activity, and rendering her general condition precarious in the extreme. The phthisical symptoms that remained dormant, under the action of minute doses of opium, became active again. The dose was increased, under the delusive idea that it would allay dangerous symptoms, until thirty grains of morphia were taken per day. She was placed under treatment, her dose reduced to fifteen grains per day, with cannabis indica two grains. The treatment as suggested for consumption was persevered in ; an improved condition was observable from the start, and she gained flesh at the rate of three pounds per week. The opium was gradually reduced, and the reduction was followed by rapid improvement of condition, until a dose of three grains a day was reached, when reaction was excited, followed by nervous excitability, coughing, and profuse expectoration. No attempt was made at a farther reduction of the opium. The system soon adapted itself to the dose, when the phthisical symptoms abated. After waiting a few days, a reduction was ordered, resulting in the return of the cough, and other symptoms denoting relaxation. During the next three months several unsuccessful attempts were made to reduce the amount below three grains, with the same effect. Then seven months were ordered to intervene, before trying again, and then an attempt in that direction was accompanied by no recurrence of phthisical symptoms.

Two years afterward she reported herself free from

symptoms usual to consumption, but unable to take less than two grains of morphia per day. Our experience justifies the assertion, that the phthisis, independent of the opium habit, could have been cured, or its activity destroyed by the treatment suggested.

CHAPTER XIV.

CASE No. 12. *Demonstrating the influence exerted by opium over cancer.* Mrs. E. E. H., forty-eight years old, married for eighteen years, but having no children, enjoyed remarkably good health previous to her forty-fifth year, when cancer of the uterus was developed. The disease was not dependent upon a diathetic influence, her progenitors being particularly free from disease of that character. When the cancer attained its supposed maximum of intensity, opium was administered hypodermically, to allay pain. The relief afforded was immediate and complete, and as a fatal termination was hourly expected, she was indulged in its constant use. As the dose was increased, she noticed the discharge became less profuse, and her general health improved. On coming under our observation, seventeen months later, she was consuming fifteen grains of morphia per day, and suffered from its direct effects. The cancer was discharging, however, but little. Believing that the action of the drug suppressed abnormal discharges, a systematic course of treatment was instituted to modify the condition of the cancer in connection with the measures adopted to relieve the patient of the opium habit, which had now become a greater curse to her than the original disease. Preparations Nos. 1 and 2 were compounded in the usual way, No. 1 on a basis of twelve grains of morphia per day. The patient experienced no disagreeable symptoms in making the change from fifteen grains administered hypodermically, to twelve grains taken by the stomach. She felt relieved, and thankful that she was able to stop the needle, as its use had excited great pain. The discharge from the cancer was very offensive, having the characteristic odor generally accompanying the disease. The following injection effectually destroyed its odor, proving a source of great relief to herself and family :

R.   Tr. Goa, ℥ i.
Chlorinated lime, ℥ i.
Aqua, ℥ v.

M. Sig. After washing the parts thoroughly with warm castile soap-water, inject one-half of the above.

We will note a case, treated subsequently for epithelioma of the lower lip, when the following preparation was applied, in the form of a paste

R.   Pulvis Goa, gr. x.
Chlorinated lime, Ɔ i.
Sanguinaria pulvis, gr. iv.
Vaseline, ℥ i.

M. Sig Cover the diseased part with a thin coating every morning, until sluffing of the diseased mass is excited, which will, through a process of suppuration, cast off the cancerous growth, leaving the healthy tissue intact. Our diagnosis was confirmed by microscopic

demonstration. The cancer had all the essential characteristics. It was, however, in its incipiency. We
consider chlorinated lime pre-eminently useful in separating morbid deposits, without infringing upon the integrity of healthy tissue, unless used strong enough to excite
its escharotic action. We mention the treatment of this
case, occurring independent of the opium habit, in the
hope that this treatment may be further tested, as the
results so far obtained by us are encouraging ; the patient reporting no return of the disease two years after
treatment Although active sluffing follows the application of the lime, no pain is excited worthy of attention. Our first patient under consideration improved in
every way after reducing her opium one-half. Her
appetite became voracious, the digestive functions acted
well, appropriating for the general system fifteen pounds
of flesh during the first forty days. Chlorinated lime,
compounded as suggested, was injected twice a week.
An injection containing :

·   R.   Salicylic acid, gr. v.
                 Glycerine, ℥ ii.
                 Aqua, ℥ ii.

M. Sig. was used once every two weeks, before retiring, after washing the parts well with warm water.
Parasitic life was destroyed by its action. We recommended as an injection, to be used every night and
morning—

℞   Borax, ℥ i.
            Chlorate potash, ℥ i.
                 Aqua, Pt. i

M. Administered internally twice a day, at nine and three o'clock.

℞.   Iod. Pot. ℥ i.
            Bi. chlo. Hyd. gr. i
                 Tr. Iodine, 3 v.
                     Aqua ℥ v.

M. Sig. One teaspoonful in a wine-glass of water.

This was given to assist in destroying, and eliminating from the system, morbid products, that possess extrotic activity when deposited, as they are then transformed into an organized, adventitious mass, having within itself the power to add to morbific accumulations, or act as a nucleus for a new growth. Constitutional treatment should be faithfully persevered in, as a cancer is a local manifestation of a constitutional disease, which accounts for the negative results attained by the knife.

Our patient's cancer was unfortunately located for well directed local treatment to be applied. We could not hope for permanent relief to follow. We insisted upon perfect cleanliness of the parts, with the constant use of injections, as suggested, combined with constitutional treatment. Reducing the opium to a point where its antiphlogistic properties could be appropriated by the system, without its debilitating effects, we surrounded her carcinoma with the primary conditions essential to a cure. Two years afterward, she was attending to household duties, consuming but two grains of morphia per day, and having a good appetite, supported by tonic remedies. She had kept her adipose tissues up to a normal standard. The tonics were changed occasionally, to avoid losing their fine medicinal influence over the system.

The discharge from the cancer being scant, and free from odor, all the extrinsic symptoms indicated that life would be prolonged many years.

CHAPTER XV.

CASE No. 13. *Showing the length of time that opium has been taken in individual cases.* Col. II. B. A., sixty years old, had taken opium, or its compounds, during thirty-nine years. A remarkable feature presented by his case, consisted in the fact that while laboring under all the decided effects of the opium, which rendered his life miserable in the extreme, he had never suffered a symptom indicative of disease. He went to his law-office every day, during that period. His practice had gradually left him many years before, owing to his being incapacitated for business by the drug. It was one of his many eccentricities, to maintain the external appearance of having a flourishing law business. His hat was full of worn-out papers, and he would have confidential talks with judges, and court-attendants.

His physiognomy presented a startling and unique appearance, attracting the attention, and exciting the wonder, of all who saw him. His eyes were sunken and expressionless, he had a dark brown complexion, similar to the color of gum opium; his skin was covered with large, warty excrescences, and indurated squamaus crusts, desquamating every two or four weeks, leaving a dry red cicatrix, emitting a peculiar odor. He conversed in a slow, labored manner, seeming to devote great thought to commonplace events, making frequent stops, becoming confused, and starting on some subject foreign to the question under consideration.

He appeared to be dazed, and uncertain upon all subjects, except that of opium; he could give, however, an intelligent and lucid account of his experience, and of the effects of the drug upon him, from the first dose to the time of speaking. His friends observed that his mental capabilities had failed rapidly, during the three years immediately preceding the beginning of treatment.

Until that time he had been very ambitious, always having some utopian scheme in hand, predestined, in his estimation, to revolutionize the business world, and flood the pockets of lucky stock-holders with millions. This peculiarity had passed away, being displaced by an apathetic state, simulating quiet dementia. During his long career as an opium-eater, his sufferings from opium-night-mare had been actually horrible. Their repulsive features were vividly described.

For years, each night, his "sleeping vision" would confine him in a subterranean tomb, presided over by some hideous monster, whose delight consisted in slowly diminishing his living grave, by bringing its walls together, until he was made to endure the agony of a thousand deaths by slow suffocation. He would endeavor to liberate himself by digging his way to a supposed surface, working with desperation born of a paralyzing fear, yet months and years would pass while he was employed at his ceaseless task. He could make no impression upon the mass to be removed; some unseen power replenished it. When he was worn out with years of labor, the watchful monster would slowly approach with stealthy step In a paroxysm of fear, the victim would make a desperate rush for liberty, and awake to find himself bathed in a cold, clammy perspiration, and discover that he had been dreaming but a few minutes.

The pleasure of finding his terrible experience to be unreal, and his thankfulness at being awake once more, was marred by a dread of again going to sleep, as the same dream would haunt his sleep, time and again, varied only by different surroundings, each more fearful if possible than the last. Ancestral ghosts would shut him up for months, denying him his usual supply of opium. He would suffer all the agony of abstaining, with the same acuteness of feeling that the reality would inflict. He got but little undisturbed rest, and suffered enough to derange the strongest mind.

During the first five years of the habit, he consumed gum opium; he then resorted to laudanum, and finally took morphia, at the rate of fifteen grains per day, during twenty years. He early discovered the necessity for controlling his desire to increase the dose, and did not allow himself to go beyond the amount named.

Nos. 1 and 2 were compounded, No. 1 on a basis of twelve grains of morphia per day, for ninety days.

Replacing No. 1 with No. 2 only every third day, we gave his system the advantage of a very gradual decrease. Warm baths of salt water were taken every night, *lubricating his face after each bath* with—

R   Acid Carbolic, gr. ii.
         Vaseline, ℥ ii.

During the first thirty days of treatment, he gained eight pounds in weight, having a good appetite. His complexion began to clear, by desquamating the old crusts, and showing no tendency to form new ones. The change produced, by reducing his dose, led to enthusiastic expressions of relief on his part; the improvement consisting chiefly in his ability to sleep quietly, free from the horrors of opium night-mare.

He continued to improve rapidly until the forty-fifth day, when symptoms of relaxation were marked by a hyperæmia, affecting the face and head, with hot flashes and vertigo, which warned us to take more time in reducing. The use of preparation No. 2 was suspended for ten days, and the unfavorable symptoms disappeared. No more trouble interfered with his progress, until the sixty-fourth day, when he fell upon the street and was found unconscious. He voided both urine and fecal matter while in that state. We arrived a few minutes later, when a hot bath revived him. He described his feelings, before losing consciousness, as peculiar. Although he recovered his former condition in a few days, his powers of recuperation seemed to have reached their limit. The system stubbornly resisted our efforts further to stimulate assimilation. Prostration, accompanied by temporary loss of mind (often observable in patients having encephaloma), constituted a prominent symptom, requiring the exercise of great caution in reducing the drug, in order that the system should adapt itself to its new mode of life. Paroxysmal attacks of nausea, with hot flashes, proved troublesome, yet the patient emphatically declared that his sufferings were heaven in comparison with what he had endured from the opium. Testing the strength of his medicine, we found that he was consuming three grains of morphia per day. A fresh No. 1 was compounded on that basis, for ninety days, and instructions were given that No 2 should be used every fifth day. Taking frequent rests in reducing the last grain, we supported the system with tonics, and succeeded in withdrawing the last infinitesimal amount of opium during eight months of treatment. The only symptom, demanding a prolonged treatment, was an abnormally rapid circulation. If the system was not sustained by a certain quantity of opium, the temperature of the body went as high as $102\frac{1}{2}°$ to $103\frac{3}{4}°$, presenting the phenomenon of fever heat, without the essential condition of fever being present. This state of things was in conflict with the hypothesis that animal heat results from the oxidation of the hydro-carbons. Our patient enjoyed remarkably good health, for two years after his cure, when he died after a sickness of eighteen hours, with pneumonia.

CHAPTER XVI.

CASE No 14. *Showing the tendency of the drug to excite melancholy.* Mrs. M. S., married for twelve years, having two children, inherited a frail constitution, which did not prove equal to the task of sustaining the physical burdens incident to marriage, and the child-bearing state. When an overdraft has been made upon the vital forces, as it had in this case, the result that inevitably follows will be progressive anæmia, or nervous asthenia.   In that condition, the system seems to demand stimulus. If this desire is indulged, a habit of using stimulants is quickly formed.   The patient now under consideration formed the opium habit while nursing her last child. The usual symptoms incident to the use of the drug were present, in an aggravated form, coupled with a state of mind bordering on acute mania, superinduced by anæmia of the brain and spinal cord.   She was denied the boon of sound, quiet sleep, owing to a continuous spasm of the nerves, exciting contractions of the muscles, denominated by opium eaters "jerking" While the habit made her nights wretched, her sufferings during those hours did not approach the misery that transformed each day into an irksome waking nightmare.

The loss of mental balance rendered life a succession of horrors.   She could not direct the mind's application.

and would dwell on one painful subject, with no power to throw off the thought, or shape its course. She had for months attended a phantom funeral, in all its mournful details; her countenance, by long association with such thoughts, had mirrored those impressions upon it. Her face had the stony, woe-begone look, so often observed in those who have buried their loved ones, and are left alone with nothing to live for, as if Pandora had willed them the contents of her box, minus the hope. She seemed to be living under the influence of an idealized horror. With great exactness she would arrange her emblems of mourning, and go and come from the church, going through all the ceremonies pertaining to the interment of the dead, consuming the same time in thinking over each detail that an actual funeral would require.

The terrible monotony of her situation was broken only by haunting fears of impending calamity. Her mind's eye would picture her husband, or child, being brought in mangled by an accident. In going out she was in constant dread of a wall or sign-board falling upon her, and she was unable to dispel such fears. She was taking McMann's elixir of opium, consuming one and a half bottle per day. Leucorrhœa and dysmenorrhœa were complications. We found, on examination, ulceration of the os uterus. To avoid prostrating our patient still further, by a prolonged treatment for ulceration with caustics, we applied the actual cautery, as we were convinced that her system needed revolutionizing completely, to excite molecular activity, her symptoms being indicative of anæmia of the spinal cord. When nervous activity is at a low ebb, in consequence of poverty of the nerve centres, it is useless to administer tonic remedies, as they cannot be appropriated by the system until the absorbents are stimulated to action through a renewal of central power. To accomplish the desired result, the actual cautery was applied, over the spinal column, from the first cervical vertebra, to the lumbra, with telling effect, giving a substantial foundation to build upon.

Bottles Nos. 1 and 2 were compounded, No. 1 on a basis of three grains of morphia per day, for sixty days; No. 2 contained one ounce of cannabis indica, as full doses of that invigorating tonic were regarded as necessary. Salt baths were taken every day, a strong attendant rubbing the extremities thoroughly. The right lumbar region was painted three times a week, with Tr. iodine. We assured the patient positively, that all mental manifestations of disease would leave her on the seventh day of treatment, never to return, unless she took opium again. Moral support should be given in such cases as this, as such means are best calculated to act upon the mind, and infuse into it the invigorating auxiliaries, hope and confidence.

Such means are absolutely necessary with patients laboring under mental aberration, to secure a normal balance, as the mind is in an uncertain and vacillating condition, and is susceptible to well directed influence, from one having the patient's confidence. The plan was successful in this instance. As the opium was decreased and the tonics increased, the patient improved with wonderful rapidity, appearing within thirty days like a different person. Traits of character and disposition peculiar to her normal condition returned with full force. She was joyous, contented, and entirely free from symptoms denoting hypochon-lriasis. Owing to great anxiety on her part to stop all medicine, she was allowed to go into the crisis on the thirty-fifth day. The result was quite a severe crisis, lasting three days. She made no complaint, however, remarking that she suffered more every day while taking the elixir, than she did now. Some leucorrhœa followed the application of the cautery to the uterus, but on examination ten days aftcrward, its parts presented a clean surface, showing that the cautery had done its work well. Electricity is not indicated in prostrated conditions connected with the opium habit, as it excites, in some incomprehensible way, nervous irritability.

CHAPTER XVII.

CASE No. 15. *Presenting a peculiar and highly interest-ing condition, rich in physiological suggestions, as the opium had completely lost its power to control nervous activity or allay pain.* Miss A. L., thirty-one years old, was born in England, of good family. She possessed a finely organized nervous temperament, was energetic but delicate, and extremely sensitive and sympathetic, with marked traits of character; yet she had no noticeable ec-centricities. She had received a classical education, fully developing her finer sensibilities, leaving her at nineteen years of age with superior attainments, suave and grace-ful, and of a commanding presence. It is not surprising that she was appointed, at twenty-eight years of age, Mother Superior of a large convent. As a child, she was ambitious and quick to learn, and was allowed to imperil her health by over-study   Many others are treading the same path, over-educating the brain at the expense of the imperfectly developed physical being, by a system-atic course of cramming, called study. Long terms are considered necessary in our city schools, extending into the spring and early summer, when the system is natur-ally relaxed, making heavy drafts upon the nervous forces. Headaches are complained of, yet do not lead to caution, or excite alarm on the part of parents. The seeds of subsequent weakness are sown, and if the over-worked pupils succeed in avoiding an early collapse of the physical forces, their parents wonder why it is that children so carefully educated should be wanting in ability to make practical application of their learning, and have so few of the resources that are essential to useful men and women.

While at school, our patient began to suffer premonitory symptoms of what afterwards proved to be an incurable nervous disease. The eyelids twitched when she was tired, she cried easily from no special cause, and tonic contractions of the muscles soon followed. At seventeen years of age her catamenia appeared, accompanied by dysmenorrhœa and a severe headache. The symptoms became more unbearable at each recurrence. Hysterical symptoms, peculiar in character, were suffered, with excitability of the erectile tissue, causing symptoms of nymphomania. Her general health was shattered beyond repair, and her sufferings each month were prodigious. They consisted in spasmodic contractions of the muscles, producing episthotnos, interspersed with clonic spasm, affecting the muscles of the arms. The body at times became rigid, yet she did not lose flesh, although prostrated for days every month, retaining no solid food during the paroxysms. She presented between the attacks a robust appearance.

After consulting, and being treated by many physicians on both continents, the opium habit was formed, closing forever the last avenue of escape. Many complications appeared during the subsequent five years. Her symptoms were intensified in severity, aggravated by the opium, making it necessary to hold her in bed, owing to the violent muscular effort. The opium having lost its power to mitigate her sufferings, chloroform was inhaled during the paroxysms. Such was her condition in coming under our observation; she was taking eighteen grains of morphia per day, hypodermically, with five ounces of chloroform. Few articles of food could be retained by the stomach, owing to its irritability, exciting extreme prostration. Eminent specialists had diagnosticated her disease as an obscure ovarian complication, *requiring removal if a cure was to be effected,* but had declared an operation under the existing circumstances to be impracticable and fraught with danger, while her system was under the influence of the opium habit.

She was bedridden two-thirds of the time, suffering intense tortures, a continual crisis being the only state to which we could compare her condition. Chloroform, although taken in large quantities, in conjunction with the morphia, failed to control the pain. Her diet consisted of weak milk punch, and beef tea. She expelled a large proportion of it by vomiting. After exhausting all of our methods, in an endeavor to have her retain our medicine by the stomach, we treated her hypodermically, withdrawing the morphia slowly, and substituting sulphate atropia. The sphincter ani participating in the general muscular irritability, would expel any substance injected, or placed in the form of a suppository, in the rectum. Remedies administered by injection failed to operate, owing to a paralyzed condition of the cutaneous surface, destroying its power of absorption. The withdrawal of a portion of her opium did not mitigate the intensity of her sufferings. Muscular contractions finally became so violent as to throw her from the bed.

Hyperæsthesia of the cutaneous surface, with cystitis appeared as complications, making it necessary to draw her urine with a catheter, causing excruciating pain. She entreated that the hypodermic injections should be stopped, as only pain followed their use. We endeavored, by administering large, and then small doses, to find if possible the amount necessary to obtain the soothing effects of the drug, but to no purpose. The opium had lost its power to affect her. Baths of any description excited convulsive effort. Nausea was now induced by the presence of either solid or liquid aliment in the stomach. During the last six weeks of her life, brandy or champagne was expelled at once, leaving us with our weapons of relief rendered useless, to witness her terrible sufferings, which beggared description. She had taken no solid food for three months without expelling it, and could retain but a small quantity of brandy, or champagne. Yet animal heat remained normal to the last.

The night previous to her death, we made a desperate effort to have her retain aliment, and met with a greater degree of success than we anticipated, although it came too late to render our patient effective service. Her

powers of life were at too low an ebb, to be susceptible
of resuscitation.   Our method consisted in a gradual introduction of maltine and brandy into the system,
through the rectum.   We used a quart cup with a faucet
and stop-cock attached near the bottom.   To the latter
five yards of French rubber tubing was fastened; to the
tubing a flexible French catheter No. 22 was attached, by
slipping the former over the latter.   The point of the
catheter to be inserted into the rectum was perforated
with small apertures, for three inches.   A small spirit
lamp was fastened by a slide to the bottom; aliment
was placed in the cup, warm, and the wick so arranged
as to maintain a uniform temperature.   The cup with
lamp attached was placed on a bracket by the side of
the bed, some three feet above its level, to obtain the
requisite pressure.   The tube was then carried under the
bed-clothing, without disturbing the patient after it was
once inserted, but allowing her to maintain any position
desired.   The catheter was passed about four inches up
the rectum, and the stop-cock turned a very little, letting
the aliment descend gradually.   As the perforations
cover a large mucous and absorbing space, food administered in this way is rapidly taken up.   If the sphincter
is thrown into contractions by its presence as a foreign
substance and endeavors to expel it, the attendant should
hold the catheter steady for a few moments, when spasmodic effort will be overcome and expulsive action stop.

If the patient is greatly prostrated, the physician need
not disturb him for hours, by removing the catheter, as

no disagreeable symptoms are excited by its presence; the flexible tubing permits the adjustment of the bed-clothing in a manner consistent with comfort.

Our patient under consideration rallied after being nourished, but reaction could not be maintained, and she died a few hours afterwards by asthenia, passing away while life's tide should yet have been flowing in; a victim to over-study in youth, exciting that which the opium completed. Repeated efforts were made to nourish her by injecting soup in small quantities every few minutes well into the rectum, but spasmodic nervous action was so great as to expel it at once. If the syringe is resorted to for that purpose, it becomes necessary to subject the patient, every time it is used, to the discomfort of changing position and to danger of taking cold, by sudden drafts of air striking the body. The sphincter will not tolerate large quantities without expelling them; the catheter allows you to introduce any amount desired.

CHAPTER XVIII.

Case No 16. *The opium habit complicated with Neuralgia.* Mrs. A. L., thirty-nine years old, had been married for fifteen years. She gave birth to two healthy children during the first four years of her married life. Her general health had been entirely good, except that she had suffered with periodical attacks of neuralgic headache, when her sufferings were very severe, the attack usually lasting from one to three days. Passing off it left a lingering sensation of tenderness over the affected side. Previous to the birth of her second babe these attacks did not appear with any degree of regularity. Her second and last labor was complicated with crural phlebitis, called also milk-leg, making her convalescence tedious, rendering her weak and anæmic. The neuralgic attacks becoming more severe, and of longer duration, opium was administered. As usual with this class of patients, its action not only made the pain bearable, but seemed to ward off the paroxysms when taken on the approach of an attack. As her case is a typical one, and this type of reflex neuralgia stands second in the array of exciting causes for the formation of the habit, we will, in connection with the history of this case, treat of neuralgia.

Our patient succeeded for several months in warding

off a large proportion of her attacks by taking one-half grain of morphia on the appearance of premonitory symptoms. The pain was allayed and sleep induced. After indulging in the drug for one year with some degree of regularity, but with no thought or intention of forming a habit, she discovered that her neuralgia was deviating from its usual course, having changed in character; the attacks coming with perfect regularity either just before, during, or immediately after, her menstrual period. The attacks now gave few premonitory signs, and began with greater violence than before, being often accompanied with nausea, obliging her to go to bed at once.

Two or three grains of morphia did not now produce the quieting and soothing effects at first obtained from one-eighth of a grain. The patient's menstrual functions lost their regularity of action. Instead of a flow of six days, as was usual before she began to take opium, she now had from one to three days only a scant secretion. The flow often stopped suddenly, and was succeeded by bloating and a feeling of oppression which indicated that the act had not relieved her.

While few premonitions immediately preceded her attacks of neuralgia, she experienced passing sensations at all times, indicating, as she believed, the approach of her dreaded enemy, and influencing her to take morphia between the regular attacks through fear of their untimely repetition. The opium had now done its work by creating changes in her system and subjecting her to pains of an uncertain character, which seemed to indicate that neuralgia was approaching, when in reality the pains were the legitimate results of previous doses of opium. The system coming out from under the influence of the drug, suffered relaxation, and simulated the pains of the disease it was taken to control.

Our patient did what nineteen out of twenty persons do in such circumstances; she took small doses of opium to ward off anticipated paroxysms, each dose making it absolutely necessary for another to follow and control the reaction from the last. When fears haunted her of evil resulting from the regular use of opium, fears suggested by progressive loss of flesh, constipation and other debilitating symptoms resulting from its use and abuse, she would make an effort to stop, but only to return to what was now necessary to her being. The discontinuance of the poison at once brought back old symptoms intensified many degrees in severity by relaxation.

After the habit had taken thorough possession of her, the neuralgia returned with its former violence, and the opium had now lost its power to mitigate the pains of the disease or modify its course. In trying to allay the pain she increased the quantity of opium rapidly, and was soon consuming sixteen grains of morphia per day.

She was weak and anæmic, appearing prematurely old, and her catemenia discharged but twice a year. Before noting the treatment of her case we will refer to the disease characterized by pain known by the modern name of neuralgia, and its treatment, independent of opium or its use. Little can be written regarding the pathology of this disease. It presents only a morbid sensibility that does not account for the severe pain induced.

To avoid confusion, and epitomize all the essentials that interest us, we will consider the disease under two separate heads, and enumerate the important or exciting causes and symptoms of each variety.

An active factor in producing sciatica and neuralgia, causing a paroxysm or an exacerbation of pain which is circumscribed and is contusive, tearing, or lacerating in character, is a reflex action and follows an exciting cause situated in the uterus, ovaries or rectum; neuralgia from this cause will continue for an indefinite time, unless removed by direct treatment, or until nature, by a slow process, corrects the abnormal condition. Women are subject to this kind of reflex influence, and suffer paroxysms during or after their catamenials, and after abortion. To administer opium and allow the exciting cause to remain, is to invite a habit; it is only to stop pain by overpowering nervous excitability, rendering the patients oblivious to painful impressions, while the exciting cause remains active.

Neuralgia appearing periodically should arouse the physician's suspicion, and suggest a local cause which must be diligently searched for There is a form of neuralgia, producing a local manifestation of pain, which is, as it were, a moan from a starving nerve, and is produced through a progressive action upon the part of the system retrograding from a normal standard by its inability to appropriate nutritious principles sufficient to replace tissue waste. This is accompanied by a want of energy, mental apathy, lack of buoyancy, and general inertia, suggesting by their combined appearances the therapeutical agencies necessary to restore the system to a normal condition. As the tendency of opium is to absorb the deposits created by the hydro-carbons, preventing further tissue change and assimilation, its use is decidedly counter indicated. Close observation demonstrates the fact that the nervous system receives its life-sustaining proportion of nutrition after all other functions necessary to life have appropriated their share.

The nerves are often the first part of the system to present symptoms of degeneration, such manifestations being generally local, caused by pressure or other obtuse influences. The nerves will be profoundly affected when the functions are generally free from disease.

The nervous system is also slow to repair any damage inflicted upon it. After it is surrounded with favorable conditions, the general system has to generate an excess of vitality to transmit force to an impoverished nerve and place it upon a normal basis. While in a weakened state, the nerves are susceptible to outward impressions, taking on irritable, painful conditions from slight causes, as sudden revulsion of feeling, cold, the night air, hunger, and all excesses causing fatigue.

The exacerbations are of variable duration, lasting from one to sixty hours. If the attack exhausts its force, as it usually does, on one special nerve, the patient can discover the trunk affected, with its ramifications, by the pain. Congestion is often present. with increased heat at the site of the pain; yet fever is seldom present, and would indicate the approach of some complication. Tenderness under pressure made with the finger tips, and especially where the nerves emerge from the skull or spinal canal, is usually present yet steady pressure often gives momentary relief.

Any act, like coughing or sneezing, that jars the body or excites sudden congestion, greatly intensifies the pain. When the ophthalmic division of the fifth nerve is affected, the pain is excruciating, accompanied with congestion, redness, and swelling of the conjunctiva. Lead poisoning excites neuralgia of a severe character.

If a person who has previously been healthy and robust is suddenly attacked with neuralgia, a thorough examination should be made of the water communications, and if canned fruit has been eaten, the empty cans should be examined, to find if lead is present in quantities sufficient to cause a paroxysm. If such be the case, the attack will be amenable to the action of bromide potassium in doses of ten grains, three times a day, after meals; sulphur baths will eliminate the lead from the system. If the attack is induced by anæmia, we must endeavor to find the first or exciting cause for that condition, and remove it by appropriate tonic treatment.

Malaria often excites neuralgia, and affects the sciatic nerve. When the attack is of that type, quinia and gelsemium, with bitter tonics, will allay the pain, by removing the cause. When the neuralgia follows progressive anæmia, the relief of pain is an important matter, calling for immediate action. We cannot hope to build the patient up at once, therefore the indications demand that we control the pain of this attack and build up our patient afterwards, rendering him strong enough to resist the disease, by exciting the assimilative processes to action.

The circumstances attending each individual case will suggest the extent to which curative measures can be carried. We were called at nine o'clock in the evening to see a lady already dressed for a reception. She was slightly built, with a highly nervous organization, a paroxysm of pain affecting the fifth pair of nerves had suddenly prostrated her. Her sufferings were intense, yet she wished to enjoy the evening by attending the party. We found her neuralgia to be complicated with gastric affections, and for the sake of temporary relief, we administered the juice of seven lemons, which produced the desired result immediately, so that she was able to participate in the pleasures of the evening. The following day we placed her under treatment for the purpose of removing the cause by improving nutrition; we put her on ten drop doses of diluted phosphoric acid after meals, with salt-water baths four times a week. Exercise, appropriately taken, and gradually increased, was insisted upon ; her bowels were regulated by one tablespoonful of olive oil containing glycerine sufficient

to flavor, taken every night, compounded with gentian
and Tr. nux vomica.

The acid can be alternated by giving every other
week the following tonic:

R. Cannabis indica fl. ext. ℥ ii.
     Iodide lime, gr. v.
    Syrup Ferri Iodo, ℥ i.
   Peruvian Comp. Tr. ℥ ii.
  Chamomile Flowers Tr. ℥ i.
M. Sig. Teaspoonful after meals.

We have often succeeded in giving relief by administering ten grain doses of chloride of ammonia The
actual cautery, at a white heat, passing the instrument
quickly over the track of the nerve affected, will prove
efficient.   As a last resort we have successfully used, when
the disease was of an inflammatory character, a hypodermic injection of turpentine superficially over the
nerve, being careful not to strike or wound that organ ;
the parts become puffy and red, but soon assume a normal appearance.   Using from 3 to 5 drops of turpentine,
to 2 of alcohol and 2 of water.

This last method of allaying the pain of neuralgia may
be denoted empirical, but we have the comforting reflection that all new applications of remedial agents
were so considered at one time.   Unilateral neuralgia,
affecting one part of the head, and sciatica often follow
uterine complications.   When such a cause is not suspected by the patient, and if no rational explanation

presents itself to account for the phenomena, a thorough examination of the uterus will often bring to light abnormal states, the pain ceasing on their correction. Suppression of the menstrual flow will excite sciatica; sciatica will also follow child-birth and abortion. Thus the great necessity for washing out the uterus with carbolized water at as high a temperature as can be borne, after it has expelled its contents, to remove all retained fragments of the placenta that would excite a paroxysm by reflex action.

If sciatica supervenes, the physician is justified in assuming that the uterus still retains irritating substances. He should proceed at once to relieve the difficulty by placing the patient in position, and with the aid of a speculum and curette emptying the organ of the offending mass; opium under the circumstances would only prolong the pain, and complicate the case. The rectum impacted with fecal matter, in conjunction with a torpid liver, will excite sciatica of a spasmodic character.

Podophyllin is indicated, with an ænemia to clear out the lower bowels. While awaiting the action of remedies, the pain can be mitigated by the application of poultices containing a liberal supply of capsicum, also by wrapping the thigh in cotton batting, drawing a bandage closely over all. Small doses of hydrobromic acid will induce sleep, or sleep may be had by inhaling the steam from Nit. Pot. If there is reason to suspect a scrofulous diathesis, from the characteristic symptoms presented by the patients themselves, their immediate family, or their offspring, the iodine potassium, with bichloride of mercury, will effect a permanent cure of the sciatica or neuralgia.

In seeking to arrive at a correct appreciation of the situation, valuable information may often be obtained by requesting the presence of the patient's family. When the patients present no evidence of scrofula, their children will often display marked symptoms of that condition by a tumefaction of the upper lip, enlargement of the head, swelling of the nose and abdomen, with flaccidity of the muscles, and the fact thus discovered enables the physician to prescribe intelligently.

The patient under consideration not only had neuralgia, but the opium habit also, so that it was necessary to treat both complications at the same time. While the opium could not control pain, as her system was habituated to its use, yet its withdrawal, or a reduction of the quantity below a certain amount, would greatly excite and intensify suffering, rendering an active treatment for its control imperative. Otherwise, the severity of the suffering would compel the patient to return to the drug, and break treatment. Preparations Nos. 1 and 2 were compounded as usual, guarding the system against any sudden changes.

As we anticipated, the patient improved wonderfully, gaining in strength and flesh during the first thirty days, and without experiencing any premonitory symptoms indicating the approach of neuralgia. Such encouraging evidence of an auspicious termination to treat-

ment can be accounted for, and should not be the basis
for forming hasty or erroneous conclusions regarding
the final result. It should rather prompt one to prepare
with special care for the crisis. The system is gradually
becoming liberated from the habitual consumption of
opium, calling into active requisition all the primary
forces necessary to a rapid appropriation of nutritive
principles, and their transformation into healthy tissue.
The circulation becomes strong and full, carrying an
abundant supply of oxygen to the periphery; returning,
the corpuscles are loaded with effete matter to be exhaled.
The absorbents, through a relaxation of their fibre, perform their part in a manner highly conducive to an increase of adipose tissue.

A harmonious action upon the part of the functions
tends to sustain a vigorous animal existence, suppressing, for the time being, painful tendencies that are the
legitimate results of weakness produced by previous indiscretions. During the crisis the system loses its
aggressiveness through a partial paralysis of the central
functions of nervous life, owing to a reaction on the
part of the economy doing without its habitual stimulant. This deprivation destroys the uniformity of nerve
supply, and causes it to spend its force in a violent and
spasmodic manner; throwing one set of nerves into an
irritable and painful state, leaving others equally important, without force sufficient to sustain a normal action.
This condition was exemplified by the patient under consideration, for when the excessive amount of opium

taken was reduced she improved in every way. but when the amount was reduced to one-half grain dose per day, relaxation ensued, followed by a severe neuralgic attack. Cannabis indica was administered in three-grain doses, with one-fifteenth of a grain of cicuta, alternating with phosphoric acid. Great relief was derived from warm salt-water baths. Finding by repeated experiments that the morphia taken was having no anti-spasmodic effect, but was only prolonging her misery by not allowing the system to return to a normal condition, we ordered it to be stopped, as there was danger of wearing out the patient's powers of enduring pain. In that case she would have felt herself justified in returning to the drug for temporary relief. The crisis continued forty-eight hours, with varied degrees of intensity, requiring constant attention to ward off spasmodic pain.

A certain comfort was secured by keeping her in an atmosphere surcharged with steam, medicated as occasion demanded. She was placed on a cot, and a rubber blanket was so adjusted as to make a closed tent, leaving an opening for pure air at the foot. A hose leading from a boiler threw steam under the improvised compartment. The leaves of hyoscyamus, belledonna and nitrate potassium were evaporated, by boiling, and used successively with good effect, allaying pain and inducing sleep. Few symptoms of irritation attended the crisis, which could not be readily controlled by the steam impregnated as described. The neuralgia was the only manifestation which showed a disposition to resist palliative measures. She declared the pain to be more bearable, however, modified as it was by treatment, than when she was taking full doses of opium. Two years subsequent to her cure of the habit she reported herself in excellent health, with a tendency to obesity, being heavier by thirty pounds than ever before.

Her menstrual functions have been regular ever since. For six or seven months following her cure premonitory symptoms of neuralgia were experienced every month, growing less pronounced at each recurrence, until they entirely disappeared. These symptoms during their continuance yielded to five drop doses of Tr. iodine in a wine-glass of water. On discharging her we recommended one lemon or two oranges before breakfast each day, and our advice was taken.

CHAPTER XIX.

CASE No. 17. *The opium habit complicated with Rheumatism.* Capt. C. M., fifty-four years old, of a lymphatic temperament, enjoyed excellent health prior to his forty-third year, when he was attacked with what afterwards proved to be subacute chronic rheumatism. The pain being intense, opium was administered to temper its severity and induce much needed sleep. As usual in chronic painful conditions, the opium seemed to be a completely efficacious remedy, but each dose paved the way for its repetition, by exciting cellure contraction, controlling pain while its direct influence lasted, only to relax the system and induce increased pain and misery, until nature had paid the full debt, with compound interest.

The effects of the opium are so pleasing and the reaction so disagreeable, that few patients who are already suffering from a painful disease will endure the additional paroxysms of pain when the means are at hand for effectually allaying them, and they turn to the opium for relief. Each reaction so far exceeds its predecessor in severity that the sufferers yield to the supposed inevitable, and continue the opium.

We anticipated no trouble in gradually depriving this patient of the opium, although it exerted a peculiar influence over his system, increasing his supply of adipose tissue instead of absorbing it.   When the drug produces a bloated condition, the appearance of the victim is more striking than the emaciation usually resulting from the habit.   The skin has a ghastly hue, closely resembling a dough face, the lips are puffy and bloodless, eyes dull and expressionless, the sexual desires are also increased, the case presenting in all respects a strange contradiction to the usual effects of the drug.

The paramount indication when the opium habit coexists with painful complications, is to overcome or control the latter by vigorous measures.   To that end we directed our efforts.   We compounded Nos. 1 and 2, No. 1 on a basis of fifteen grains of morphia per day, as the patient was consuming twenty grains per day, and ordered the replacing of No. 1 with No. 2 every third day.

As Rheumatism is one of the diseases for which opium is given for the sake of its soothing or soporfic effects, we will give a few suggestions concerning its therapeutical requirements.   The pain of rheumatism is the manifestation of a systemic condition, pre-existing in the blood, supposed to be caused or excited to activity by various extrinsic influences, such as exposure to cold, a humid miasmatic atmosphere being enumerated among its causes.   We cannot appreciate the force of such reasoning, or accept the theory; when disease overtakes us we are apt to seek for and endeavor to find some special cause, through ignorance of its true pathogeny; and we easily fix upon some supposed exposure as a cause, when in fact the exposure or indiscretion was purely an accidental coincidence. If a congenital or inherited predisposition exists, and is on the point of becoming active, a sudden exposure without doubt develops it into an active factor. The first cause may be present in the economy in a dormant state, not sufficiently strong ever to become active so long as it is rendered innocuous by a harmonious working of all the functions. It can only be awakened to activity by decided changes taking place in the system which conspire to produce that result, such, for example, as the accumulation of effete matter, when from any cause, waste matters are not eliminated from the system; the functions of the skin are not normally active, allowing the system to accumulate matter that is morbific in its tendencies.

Associated with such surroundings the rheumatic tendency becomes active and developed. When it pre-exists in a decided measure, slight causes will render it active. In the case of young women who have developed rapidly, and are passing through a climacteric period incident to early life, when the menstrual functions are becoming active, a slight indiscretion will precipitate an attack that is liable, if not overcome by active measures of treatment, to terminate in cardiac complications.

If they pass that interesting period without an attack, a very small proportion of women will ever suffer from the disease in after life. There are four classes of subjects who are attacked by the disease, and the fallacy of applying any one single remedy to every case is only too apparent. Nowithstanding the fact that the disease pursues a uniform course, its origin and type are as different as the constitutional states and conditions presented. If ultimate success is to be attained, each condition must be fully recognized, to treat it intelligently, by stripping it of everything that is adventitious.

We have presented for treatment acute articular rheumatism, articular, subacute, and chronic, gonorrhœal, syphilitic and muscular.

Besides young and growing persons those especially liable to rheumatism are the feeble and anæmic, the obese and apparently strong, and others who are in perfect health, but who have an inherited diathesis, and the drinkers of malt liquors ; yet we have no data on which positively to base this latter assertion. Pathologists have shown that a morbid principle exists in the blood as a materis morbi, in the form of lactic acid, and uric acid in gout. Admitting that such is the case, experience has amply demonstrated that we can not in all cases accept that fact as a basis for rational treatment and expect success to crown our efforts. Therapeutics having for their foundation theoretical pathology, naturally lead to routine treatment, and the legitimate ultimation of the latter has been the traditional advice of "Six weeks and Blankets," which does not tend to inspire confidence in the learned profession.

The disease, exciting as it does peripheral paralysis which closes the cutaneous excretions, we should not feel surprised if morbific matters accumulate in sufficient quantities to admit of demonstration. The broad indications suggested by the condition of the patient are to improve the weakened circulation, eliminate morbific principles, and give tone to the general system.

Rheumatism of the young, if treated in time, is usually manageable. With this class the forming stage is brief; a chilly feeling precedes the febrile attack; the skin and adipose tissue over the extremities appear shrunken and cold; the tongue is coated, and the breath fetid. Loss of appetite and constipation are both present, in a large majority of cases. Darting pains are complained of, extending from the heart down the spine, to one or more of the articulations. If the disease is allowed to take its course, fever succeeds the chill, and the pain becomes intense in the joints affected, accompanied by redness and swelling, making it necessary to keep the part perfectly quiet, the pain being insupportable on the slightest movement; even jarring the bed excites severe pain.

When constipation is present, an enema of soap and water should be given; after the lower bowels have been emptied, the following preparation may be injected and retained:

Castor oil, ℥ i.

Podophyllin ¼ gr.

A soda bath will often break the attack, and its use

should not be neglected; one pound of soda can be used with warm water sufficient to cover the body. The patient may then be placed in a warm bran-pack, an attendant applying the warm mash, and holding it in place with rubber blankets. Jaborandi, in teaspoonful doses, may be taken every half-hour, until a profuse perspiration is excited. On removing the patient from the pack, he should be well rubbed with towels and put to bed. Three hours afterward half-drachm doses of the tincture of chloride of iron can be administered, well diluted, and combined with Tr. capsicum. This treatment often cuts short the duration of the disease.

Patients whose sufferings arise from an inherited tendency, can be treated with a view to destroy the morbific first cause. It is well to subject them to the action of warm packs and juborandi, to stimulate cutaneous activity, but it is necessary to follow this at once with vigorous constitutional measures, such as the administration of iodide of potassium with Tr. iodine, and bi-chloride of mercury, in full doses, regulating the bowels with enemas of oil every night. We have relieved patients with the iodide thus combined with remarkable promptitude. Large doses are absolutely necessary, and its use must be continued for some weeks after an attack, to ward off subsequent paroxysms by removing the cause. If this treatment fails, salicylic acid, in scruple doses, administered with an alkaline carbonate, will often work wonders, a cure being effected in a few hours. The medicine should be compounded each day, as chemical changes take place when it is put in solution as above directed, and when mixed in glycerine and water it is not so effective.

While the above mentioned effects can be relied upon in many cases, this prescription often depresses the vital forces, rendering the patient liable to sudden relapses. This is the case especially when the preparation is administered to anæmic patients, or those who have functional or organic diseases of the heart, the treatment often depressing that organ to an alarming degree, and causing great irregularity in its pulsation. Flabby subjects presenting a delusive appearance of good health, who have lowered their vital forces gradually by consuming malt liquors, and who suffer greatly from indigestion, have a weak circulation, and when attacked with rheumatism they are quickly prostrated, and generally have endo or exo-cardial trouble. They should be given a preparatory treatment with packs, followed by full doses of the alkaline carbonates. The juice of seven lemons taken every other day for a week, will give good results. The bran-packs are applicable to all cases, and should be vigorously persevered in. To maintain immobility of the affected joints so as to avoid the pain that would otherwise be inflicted in moving them, the physician should apply starch or plaster bandages, putting a rubber bandage over the plaster. For the relief of pain, and to induce sleep, the hypodermic use of atropia has proved highly efficacious. The general system should be built up as soon after an attack as possible. Quinia and the bitter tonics can be depended upon, combined with cannabis indica, to accomplish the desired end.

When from neglect in calling a physician, and endo-cardial inflammation supervenes, the physician should administer at once, carbonate of ammonia in solution of the acetate, in three-grain doses, to check the inflammation, and secure a solution of the fibrinous exudation, which impedes the action of the heart by its presence. Digitilis will give tone to, and control its action.

Our patient having rheumatism, complicated with the opium habit, improved progressively under the needed influence of reduced doses of opium, and experienced no rheumatic pains until his dose consisted of three grains per day, when the rheumatism made its appearance. As it was chronic, we did not stop reducing his dose, but treated the disease with pack, iodide potassium, and mercury, and succeeded in bringing him through both with but little pain.

We subsequently relieved a physician of the opium habit after twenty years' habitual use. The opium was originally taken to control the pain of chronic rheumatism. He suffered from the disease nearly every winter during that time, but had sustained no paroxysms for some months ; on his coming under treatment for the habit, we expected that vigorous measures would be necessary to allay the intensity of the rheumatic pains when relaxation should set in. But we were agreeably surprised to find that there was no recurrence of the disease.

CHAPTER XX.

OPIUM SMOKING.

ALTHOUGH our experience during the last twelve years with those addicted to the opium habit has been extensive, we having observed and treated between three and four hundred persons who consumed the drug or its compounds in every conceivable way—persons representing all the professions and nearly every trade—yet we have been called upon by only three men who wished to be relieved of the habit of smoking opium. One was a christianized Chinaman, and two were our own countrymen. The latter did not smoke the opium according to the method approved by the Chinese, but dropped small pieces of gum opium into their pipes after getting the tobacco contained therein on fire. By so doing, they did not receive the full constitutional effects of the drug, because of a lack of heat, and the presence of tobacco modified its ultimate results.

They felt the need of its stimulating effects four or five times per day, while the Chinaman's system was more thoroughly sustained by indulging once a day. He smoked then until he was stupefied and slept from a half-hour to one hour each time. The Americans would stop smoking when a drowsy sensation was experienced. The *modus operandi* of smoking the opium peculiar to the Chinese does not allow any of its substance to escape. A pipe having a large straight stem is used, the bowl being small at the bottom, coming to a point so that it can be screwed into the stem. The bowl gradually enlarges as it approaches the top, being similar in shape to that of an old-fashioned clay pipe ; a spirit-lamp stands by the couch used by the devotee in sleeping off the effects. The stem is taken in the mouth and a reclining position assumed ; an attendant then places a small lump of gum opium on the end of a wire, and quickly exposes it to the blaze of the lamp, twirling it dexterously in the meantime to warm it on all sides. This lump is then forced to the bottom of the bowl and the wire withdrawn, leaving a small air hole through to the stem.

The flame of the spirit lamp is now allowed to touch the opium ; at the same time the smoker takes a long pull, inhaling the blue smoke into the lungs and expelling it through the nostrils, receiving its full effects. One or two inhalations narcotize the new beginner, causing him to experience strange sensations of delight. The memory is excessively stimulated, past events float before the mind's eye, exaggerated and changed, presenting varied forms, beautiful as they are strange. Erotic thoughts are wonderfully intensified, unalloyed by any desire to gratify the grosser sexual passions. Worldly anxieties and cares are effectually banished ; sleep of a peculiar type soon follows, simulating coma-vigial ; but, although sleeping, the smoker is conscious of pleasing sensations that exalt the finer sensibilities. The effects of smoking opium upon the system are the reverse, in some respects, of what would be expected ; it excites the peripheral nerve, disturbing the central functions but little, and blunts the mental capabilities, if excesses be indulged in ; yet its action is not decided. The reaction is attended by a disagreeable sensation of hebitude of the brain, with a desire, in a few hours, for a repetition of the debauch. There is no mental depression, demanding its repetition, but purely a physical craving which gives the sufferer no rest, until the fibres are again constricted.

The general strength and tonicity of the system show the same progressive loss for the want of nutrition that is suffered when the opium is taken by the stomach. Secretions are suppressed, constipation being present, demonstrating that the action of the drug in many respects, is modified but little by smoking it. We founded our treatment on the supposition that few differences existed, arising from the manner of consuming the drug, and the favorable results attained proved this assumption to be correct. The Chinaman was smoking twenty grains of gum-opium per day. Preparations Nos. 1 and 2 were compounded, No. 1 on a basis of one and a half grains of morphia daily for sixty days, that amount sustaining his system perfectly. No. 1 was reduced by No. 2 every other day. One of the Americans was consuming one drachm of gum-opium during the same time, but, as already explained, he did not receive its full effects. His

No. 1 was compounded on a basis of three grains of morphia per day, and this proved to be sufficient to allay irritability.

A troublesome symptom, affecting all three of these patients, consisted in a profuse secretion from the schneiderian membranes, lining the nasal fossue; we ordered them to snuff up the nostrils weak Tr. belladonna, which controlled the excessive flow in a measure. The crisis with each of them yielded to the same methods of treatment used in cases in which the drug was taken hypodermically, or by the stomach, presenting no special differences. Their improvement was as marked, and the relief afforded by a cure as agreeable as with patients who have formed the habit in other ways.

CHAPTER XXI.

PAREGORIC, BLACK DROPS, SOOTHING AND PATENT COUGH SYRUPS.

THE custom of administering such remedies as those named above, containing as they do a large percentage of opium, is fraught with danger, and exert at all times a pernicious influence. During the second summer of a child's life, their use seems to be required, but this is an error, as they can be very well dispensed with. Their action materially interferes with tissue changes necessary to the child's nutrition and growth. When such preparations are administered to control disorders of the stomach, or to induce sleep, abnormal conditions are created which did not exist before, and which seem to demand the continued use of such drugs. Above all, such drugs should never be left in the hands of nurses, and its administration confided to their judgment, as their own personal comfort suggests large and continued doses, in order to keep the child quiet. God alone knows the number of children sacrificed yearly to this practice, or subjected to injury which acts as the exciting cause for other diseases.

Intelligent parents, who buy soothing syrups by the dozen bottles for their babes, are apt to blame their maker for the legitimate results of their own folly.

Mothers nursing babes who are cutting teeth, should correct their own indiscretions and irregularities of habit and note the salutary effects produced before dosing their children.   If a child's bowels are relaxed from inflammatory causes, induced by a morbific principle in the milk, a few doses of iodide potassim taken by the mother will act through her milk, and correct the babe's condition.   Simple diarrhœa can be checked by giving the white of an egg beaten up and sprinkled with an infinitesimal amount of white sugar.   If the color and frequency of the actions indicate a tendency towards cholera infantum, Tr. lobelia will correct the action of the liver, warding off or curing the attack.   Geranium, with syrup of rubus villosus, and compound aromatic spices, may be depended upon as a trustworthy astringent.   Tr. ginger, cubebs, and anise seed, will generally allay wind colic, while keeping the abdomen at a uniform temperature, by the use of a flannel bandage, will often remove the cause.   Crying is a physiological act, salutary in its effects, and should not prompt the mother to administer opium.   Women should avoid overdressing babes, and should expose them during the day to light and air, and thus secure the primary conditions necessary to produce sleep.   We refer to this matter because it is directly connected with our subject.

CHAPTER XXII.

OPIUM POISONING, OR NARCOTISM.

OPIUM, or morphia, is sometimes taken by mistake, or with suicidal intent. Morphia and the salts of quinia being similar in appearance, the former is sometimes taken by mistake for the latter. Persons forming the opium habit become so familiar with the drug, that a contempt for its dangers is entertained, and they sometimes poison themselves in consequence. The physiological action of opium upon the spinal nerves is direct and forcible, paralyzing motor activity, causing a thickening of the solid constituents of the blood, threatening a separation of those elements from its fluid state, and tending to cause the blood to coagulate in the capillary vessels of the brain, and thus to produce convulsions and death.

The red blood corpuscles cannot carry sufficient oxygen; muscular force is impaired; the pupils of the eyes are contracted; the conjunctile membranes are congested and dry; the mouth is hot and parched, and the pulse is full and strong at first, only to become light and quick. If profound stupor or insensibility has set in, the breathing is stertorous, the internal secretions are closed, with relaxation and activity of the cutaneous organs, suggesting the use of belladonna as the physiological antidote; and belladonna has been so considered owing to its action being the reverse of that of opium in this respect. The indications demand that the patient be immediately aroused, and the constituent elements of the blood restored to a normal condition, by the addition of liquids to arrest the accumulation of effete solid matter. To allow patients to sleep in this condition is to court death. We must place them in a large, airy room, divesting them of all clothing. Supported by two assistants, they must be forced to walk briskly by a severe flagellation with a wet towel. A full dose of ipecac should be given, followed by a teaspoonful of mustard in a half-tumbler of warm water. Liberal libations of salt and water must be taken to facilitate the action of the emetic. After the desired result has been accomplished the patient should be placed at once in a bath, and on leaving it a large enema should be given to clear out the lower bowels. After the bowels have moved, an enema to be retained, consisting of water and lager beer, should be taken. Fluid extract of java coffee can now be administered hypodermically. When the stomach has been emptied by vomiting, hot coffee and lager beer should be given alternately, a table-spoonful of each every five minutes. The latter counter-acts the effects of opium, as no other remedy known to us does, also slightly stimulating nervous reflex action, which is so liable to fail while the patient labors under the influence of opium.

The person under treatment must be kept walking,

the wet towel well laid on, being often necessary. Hot beer and coffee, by the stomach and in enema, should be persevered in, allowing only time sufficient for their absorption, before repeating the dose. The bath every twenty minutes is highly important, as it gives the blood the benefit of the water absorbed. The patient may be kept awake during the bath by tickling the feet. The bladder must be relieved with the catheter, it being impossible to discharge its contents naturally. Tr. belladonna can now be administered, but we prefer not to use atropia hypodermically, as is recommended by many physicians.

We were in consultation where a drug clerk had taken five grains of morphia with suicidal intent. The stomach had been successfully emptied, and he was apparently doing well, when a subcutaneous injection of the one hundredth of a grain of atropia was given. He fell back a corpse within five minutes. Paralysis of the muscles presiding over the heart may have been the cause, but the facts appeared too suggestive to be accounted for as a mere coincidence.

We favor under all circumstances a conservative course of treatment, and like to let well enough alone, giving the smallest amount of medicine possible consistent with the end to be attained. Active measures of treatment must be kept up for twelve hours before allowing the patient to sleep, as there is a tendency towards congestion and failure of muscular effort for that time. After the critical period is passed, and the time has come to permit sleep, the patient should be aroused, and a table-spoonful of hot soup, containing capsicum, given every twenty minutes for two or three hours. A full dose of elaterian or podophyllin can be advantageously administered to excite the dormant secretions.

The pulse is not a safe criterion of judgment in all cases, being sometimes slow and full to the last, and furnishing no premonitory evidence of coming death. In fatal cases respiration is slow and stertorous, at times scarcely perceptible. The complete or nearly complete suspension of deglutition, contraction of the pupils of the eyes, paleness and cadaverousness of the face, and convulsions followed by coma, are usually the symptoms denoting the approach of death, and should warn the physician to redouble his efforts, giving medicine by enema. A large flexible catheter can be passed down the æsophagus, enabling us by the aid of a syringe to inject into the stomach beer and coffee charged with capsicum, and belladonna ; the latter is indicated when excessive paleness is present. Its administration should be continued until its action is shown in the face, which, under its influence, will become hyperæmic. Mustard can be spread on roller bandages and wrapped around the extremities. Hot salt rubbed violently over the surface, will keep up the cutaneous circulation. The appearances generally observed after death give no definite or satisfactory idea of what the person's condition was preceding the event, as post-mortem changes necessarily confuse and mislead us. The convolutions of the brain are flattened, with an accumulation of serous liquid under the arachnoid. The vessels of the cerebro-spinal axis, the lungs, heart, liver and spleen are distended with dark fluid blo

CHAPTER XXIII.

DIFFERENTIAL POINTS IN DIAGNOSTICATING BETWEEN
OPIUM NARCOTISM, ALCOHOLISM, AND THE MANIA OF AP-
PROACHING DISEASES.—ESSENTIAL CHARACTERISTICS
PERTAINING TO EACH.

THE young physician is often called to attend a
patient, who has excited the alarm of his friends by
strange actions and peculiar symptoms. Women especially, if unaccustomed to stimulating drink, will be
hysterically delirious after imbibing, and we cannot detect
the odor of liquor on their breath. If they are under the
influence of liquor, they invariably mutter and cry, at
the same time. The lachrymal, nasal, and salivary secretions are profuse, the face red and swollen, the pulse
rate quickened, but full and bounding, respiration heavy
and regular, the eyes presenting a vacant, congested appearance, with pupils approximating the normal size.
If this same degree of delirium or mania is present, and
is excited by the approach of some acute disease, the
secretion of tears and saliva will be lacking, the eyes
being bright yet with no appearance of intelligent recog-
nition in their depth. Persons so affected speak dis-
tinctly, instead of in broken mutterings; the tongue is
coated, the skin hot, dry and harsh to the touch; the
pulse is wiry and quick, the respiration shallow and

light; the temperature elevated from one to four degrees, varying according to the nature of the disease. In opium narcotism the skin is moist, the pupils contracted, the secretions suppressed for the time being, the predominating desire being to go to sleep; the pulse is regular and full, with a normal or slightly increased temperature. Although persons thus affected are highly stimulated, their actions do not simulate drunkenness. They can talk fluently with no impediment of speech, but they are apt to reveal their condition by gross exaggerations and unreasonable propositions. It is impossible to anger or excite them, as they enjoy a self-satisfied state of mind, not sensitive to outward impressions. If concussion, compression or paralysis be present, the physician can diagnosticate by exclusion if no history can be obtained, and arrive at a correct conclusion.

CHAPTER XXIV.

THE HYPODERMIC ADMINISTRATION OF MORPHIA.

IT is not good practice to administer morphia hypodermically, except under circumstances requiring great haste and prompt action. Such urgent necessities are seldom presented to the physician in general practice. An important ground for our position is, that a mistaken idea exists regarding the dose required, many supposing that half the quantity will answer the purpose if given hypodermically. Long experience with the drug does not sustain this theory.

If one-eighth of a grain of morphia is indicated, as decided results will follow the administration of that amount by the stomach as the hypodermic injection would give. The fact that more time is consumed in its action constitutes the only difference. Again, in eighteen out of twenty cases, when opium is indicated, we can administer dover's powders or codia, by compounding in capsules ; and can add an excitant to the secretions, and ward off after-effects that are not desired. If the hypodermic injection is used, the physician is obliged to confine himself to morphia. If nausea follows its administration by the stomach, capsicum, belladonna, or quinia in minute doses will not only allay the symptom, but facilitate the action of the codia, dover's powders, or morphia, and protect the patient against suddenly developed prostration, threatening paralysis of the respiratory muscles. Morphia or laudanum, compounded with carbolic acid and glycerine, administered by enema, will give prompt results, and but seldom excite nausea.

By such methods it is possible to avoid the danger of creating obstinate sluffing ulcers, which often follow the pertusions made by the needle. It is not an uncommon thing to strike some small vein or artery by accident, and when such an accident occurs, the patient is apt to be profoundly affected and in danger of collapse. The vaso-motor nerves receive a shock; a chill passes through the economy, followed by alternate congestion and pallor. The patient and his friends present are greatly frightened and agitated over the physician's apparent want of skill. Fortunately for all concerned, these symptoms in most instances soon pass away; such favorable results do not always follow, however, as many cases are on record in which the patients became deadly pale, with set jaws, head drawn back, and respiration almost suspended, requiring from fifteen to twenty minutes of vigorous treatment to resuscitate them.

The indications, when the accident does happen, include alcoholic stimulants administered by the stomach and in the form of enemas, the fumes of aqua ammonia, with friction and warm baths. A melancholy result sometimes follows the occasional use of the hypodermic syringe, the patients soon becoming familiar with the drug and its pleasing effects. They acquire all necessary information and purchase an instrument, and soon become confirmed in the habit of using opium.

There is no drug in the pharmacopœia that so commands our admiration for its varied effects upon the system as opium does. No drug requires more judgment and skill in its administration. It is capable in the hands of the judicious physician of doing an inestimable amount of good. It can also weave a chain of endless misery, slavish degradation, and death, that can not be fitly described.

When a physician prescribes a drug so fraught with danger to the after-life of his patient, he should neglect no precautionary measures for the protection of his patient.

CHAPTER XXV.

ALCOHOLISM.

ACCORDING to the nomenclature adopted by universal usage upon the part of the profession, the term alcohol includes all beverages containing alcohol. Our province is to treat of a pathological condition created by its abuse. In prescribing for its results the physician should estimate the amount of alcohol usually contained in the drink taken by the patient and vary the strength and amount of combatting remedies accordingly. As a rule the habitual consumer of alcoholic stimulants will take the strongest compound attainable, such as whiskey, brandy, gin, or rum, adding a liberal allowance of beer. It is a well beaten path that leads the beginner from his first drink of beer or wine to whiskey.

We have presented for our consideration the periodical drunkard, who surprises his friends by going on a protracted "spree" after having lived a life of sobriety for from one to three years; he will suffer reaction and mental torment for his folly and again settle down to a sober life. A second class take their early morning dram, and will not think of repeating or increasing the amount. A third class begin daily indulgences in the morning and imbibe many times during the day without exciting intoxication. A fourth class indulge every day to some extent, becoming grossly inebriated twice or three times a month. A fifth class indulge but seldom in liquor of any kind, having no desire for it; if they take one glass, however, all will power is lost, and they consume glass after glass in quick succession until stupefied. After sleeping off the effects of their potations they require no more liquor for the time being, and may go for weeks without repeating the debauch. The tendency is to repeat, however; more time and liquor are required on each occasion to produce the desired effect, and less time intervenes between the indulgences, until finally the drunkenness becomes continual, constituting a pronounced type of dipsomania. The sixth and last condition to be considered is dipsomania, the terrors of which should excite our profoundest commiseration for its victims.

Dipsomaniacs are cursed with a craving, gnawing appetite which they cannot control or satisfy. This appetite has been cultivated to a certain extent, we do not doubt, but in a great many instances there must have been a constitutional first cause, an inherited predisposition, which, when aroused to activity, was transformed into an irresistible, controlling influence.

It is not civilized man's natural disposition to relinquish all that serves to make life happy, in order to satisfy his appetite for drink. Such is the case, however, with dipsomaniacs. Their energies and talents are directed towards the gratification of their all-absorbing desire for drink.

The several types of alcoholism described above have

varying and peculiar complications, and require for their relief a thorough knowledge and appreciation of the diseased state incident to each.

We will consider them in the order already adopted. The periodical drunkard presents no decided pathological changes, as he indulges but seldom, and allows the system ample time in which to recuperate by eliminating the poison before it has an opportunity to cause tissue changes. Although such men go to terrible excesses, in many instances prolonging the debauch for from one to three weeks, they either have a voracious appetite and consume a prodigious quantity of food, or else they depend entirely upon the liquor for aliment.

When the appetite for food is only equalled in intensity by that for drink, a protecting influence is exercised by its presence; it keeps the gastric functions busy, utilizing to a certain extent the toxic effects of the liquor, and having a salutary influence over the general economy by equalizing the circulation. It is noticeable that delirium tremens is seldom suffered by patients having this peculiarity. When the desire for food is completely suppressed by the liquor, the system is left at the mercy of the alcohol, in an unprotected state, that invites delirium, as the nervous system receives the full effects of the stimulant.

The indications for treatment in the cases of periodical drunkards are to put an end to the debauch at once by destroying the desire for stimulation, allowing the patient to become sober and conscious of his condition. This can be effectually accomplished by some friend, acting under the instructions of the family physician, administering, surreptitiously if necessary, fifteen drops of extract of ipecac in the patient's liquor; as the dose takes effect the drunken man becomes tractable and consents to return to his home. The ipecac not only vomits the patient freely, but causes a general reduction of the excitement by stimulating cutaneous activity, so that the man rapidly becomes sober.

Under pretence of relieving him, a teaspoonful of mustard in a half tumbler of tepid water, can be given to empty the stomach of its contents. When the patient is sober enough to follow instructions, a hot bath should be given at once, followed by a large soap and water enema, thoroughly to wash out the lower bowels. A capsule containing the following preparation can now be given:

$\qquad$ ℞.   Podophyllin, gr. ¼.

$\qquad\qquad$ Capsicum, gr. i.

$\qquad$ Pulvis Dover's, gr. v.

One hour afterward it is necessary to administer thirty grains of bromide of sodium. The desire for drink will be greatly or entirely allayed. Sleep is usually produced by the combined action of ipecac, baths and sodium. If exhaustion follows the reaction superinduced by the medicine, aromatic spirits ammonia, capsicum and nourishment must be depended upon. The following preparation can be administered for a few days, to control any lingering desire for liquor that may exist:

℞. Cinchona rubra, fl. ext. ℥ iii.

Ginger Tr. ℥ i.

Capsicum, ℥ i.

M. Sig. Teaspoonful every three hours.

When patients pursue their wayward course for one or more weeks and outraged nature refuses to tolerate any further abuse, and the physician finds them bed-ridden and suffering with acute alcoholism, the cutaneous surface generally presents peculiarities of diagnostic value which suggest the extent and activity of therapeutical measures to be adopted. An unequal temperature, coldness of the feet and hands, with increased heat of the trunk, are symptoms indicative of central irritation, and precursors of delirium. The tongue is coated, the pulse quick, full and bounding. The skin is hot and dry, with a puffy, congested look; the eyes are expressionless and congested, the bowels constipated, gastritis being generally present in a sub-acute form.

If the patient is a woman, vaginitis will invariably ensue as a legitimate result of her excesses, and local treatment without recognizing and removing the exciting cause, will prove utterly inadequate. Neighboring parts become involved in aggravated cases, the inflammation spreading by continuity. It is not improbable that she will suffer endometritis, ovaritis, perimetritis, urethritis, cystitis, or nephritis.

Nervous prostration following her prolonged debauch is usually excessive, with irritability of the nerves presiding over the gastric functions, causing incessant vomiting; there is also more or less hepatic congestion.

Treatment can be appropriately begun by meeting the first indication; exciting reaction with a warm bath, relieving the lower bowels of their contents at the same time with a large enema. A suppository containing—

℞   Podophyllin, gr.$\frac{1}{4}$.

Aloes, gr. i.

Gamboge, gr. i.

can be inserted as high up as possible. If a mustard plaster placed over the epigastrium does not stop the vomiting, twenty drops of chloroform mixed with the white of one egg will accomplish that result. After the vomiting is subdued cinchona rubra with Tr. ginger can be given to control nervous excitability. If the skin remains hot and dry after the bath, jaborandi in teaspoonful doses every twenty minutes will excite cutaneous activity by the time three doses have been taken.

Persistent vomiting can often be controlled on its return by applying ice bags to the spine. It is well also to vesicate the skin over the epigastric region, and sprinkle quinia on the part prepared.

Ice dropped into champagne taken ad-libitum will often act promptly. After the bowels have moved freely and the liver has become active, nervous irritability can be controlled by continuing the bath, and administering cinchona rubra as heretofore recommended. Complications can be treated as if presented under any other circumstances.

CHAPTER XXVI.

THE habit of taking a drink of liquor before breakfast, or on retiring, has been common for many years. As a custom its history is intimately associated with the early pioneer life of our progenitors, when a supposed necessity first suggested its use.

This custom is followed by a certain class of persons not especially given to excesses who are influenced by custom. Since the dangers of alcoholic stimulation have become more thoroughly understood, the habit as a custom is becoming obsolete, and is now practiced only by persons who wish to stimulate their appetites. Such drinking is no longer regarded, as it once was, as a matter of social obligation.

The only rational explanation of the small amount of misery and ill-health engendered by such a custom is that the liquor formerly used was distilled at home. The vile decoctions now sold as pure old rye or corn whiskies contain potato spirit, strychnia, tobacco and sulphuric acid.

The social conditions with which we are now familiar did not exist at that time, and some of these conditions have decided tendencies towards excess. Our forefathers took their morning or evening dram with comparative safety, not thinking of further indulgence.

If changes of an injurious character were produced, they were not noticeable, the reverse appearing to be the result, the dram-drinking seeming to have a salutary effect, and to be conducive to longevity. That this was the actual effect cannot be definitely shown. Indulgence, in such a habit in our time produces no uncertain or doubtful results. The ill effects of the practice are so palpably injurious that few medical men can be found who are willing to hazard an opinion favorable to the practice.

The lesions superinduced by the habit, are structural diseases of the kidneys, with impairment of their capacity to eliminate effete matter.

Cirrhosis of the liver is induced, and atrophy of the muscles, with pigmented deposits beneath the skin, is frequent. The extremities are poorly nourished, furnishing the primary condition favorable to the formation of ulcers, that obstinately resist treatment. These results are likely to follow when the habit is formed in early or middle life, when the system is enjoying its full vigor, and does not require alcoholic stimulants to assist it.

We freely admit that the conditions and results are different, when men of sixty years, who have always abstained from liquor, begin at that age to use brandy moderately as a tonic and aliment, as at that age the powers begin to fail, and nutrition is not sufficient to repair waste perfectly, as it does in the flush of mature manhood.

The brandy in many cases reawakens the dormant energies, by quickening the assimilative process. We would not, however, feel justified in recommending its continued use, except in special cases. Tonics and favorable hygienic treatment are usually preferable. The administration of brandy in low typhoid states, or as an aliment when progressive waste of tissue is impoverishing the system, is necessary of course.

Persons in the prime of life taking liquor regularly on an empty stomach, may persuade themselves into the belief that no detrimental results will follow, as they are free from disagreeable symptoms. Their systems seem to demand the stimulant. Many persons are convinced that marked benefits are derived from its use. A more deceptive or fallacious idea could not be entertained.

The system in such cases, from long abuse, has become changed so that it demands its usual quantity of stimulants to sustain an abnormal standard created by the liquor.

If such persons discontinue its use the appetite fails; dyspeptic symptoms appear, with pain in the right hypochondriac region; a slight fever in the middle of the day is suffered, palpitation, owing to an unequal circulation, is generally present, with constipation and nervous irritability; they present a jaundiced appearance, and lose flesh progressively. If they have the will-power to carry out their determination to abstain from liquor, nature will right herself in a few weeks by eliminating the accumulated poison; then their improvement,

ment, and is liable to do harm, by shocking the nervous
system. The family physician can modify the following treatment as the circumstances of each case may
demand. The principal indications suggest tonic and
supportive measures, with baths and a substantial diet.
To control the desire for liquor, and allay nervous irritability, the following prescription will give satisfactory
results:

℞.   Cinchona rubra, fl. ext. ℥ iv.
          Coca leaves, fl. ext. ℥ iii.
          Tr. ginger, ℥ i.
          Tr. capsicum, ℈ ii.

M. Sig. One teaspoonful as needed. The first dose
should be given at the hour at which the patient has
been accustomed to take liquor. During the first four days
four doses per day will sustain the system. The following pill may be taken at bed-time, when torpidity of
the bowels demands its exhibition.

℞.   Leptandrin, gr. x.
          Hydrastin, gr. iii.
          Podophyllin, gr. ii.
          Hyoscyamus, gr. i.
          Nux Vomica, gr. ii.
M. Pills xx.

The above treatment is also adapted to the every
day drinker, who has been accustomed to drink several
times per day, but has avoided a condition of intoxication.

CHAPTER XXVII.

DIPSOMANIA.

THE condition which we call dipsomania is excited by repeated and continued indulgence in alcoholic stimulants; it consists in an uncontrollable desire to remain under the influence of liquor, and involves, of course, the utter incapacity of the patient to attend to any of life's duties.

With hope and the finer sensibilities crushed, the victims of dipsomania are only capable of causing tears and shame, and of compromising every one connected with them. The duty of the physician in the premises is plain. He should persevere in the treatment that recommends itself to his judgment, and stay this torrent of misery that is flowing relentlessly over its bed of human hearts.

The pathological condition constituting a confirmed dipsomaniac state is well marked, demonstrating that dipsomania is a natural reflex result. Its degree of intensity is in proportion to the changes wrought in the system by the excessive use of alcohol. Dipsomania is not the result of indulgence continued for a brief period, as the system is slow to respond in this way to alcohol.

The beginner takes his first alcoholic drink as a medicine, or in a social way, and there is no appetite demanding its immediate repetition, although the effects are pleasing. The use of liquor is continued as a social habit, or for the sake of pleasing after-results, and the practice at this stage is not in any way connected with a demand from the system.

By repeated indulgence, the normal integrity of the economy is impaired, and an abnormal state is created. The system now makes imperative demands for continued stimulation, and its expression is through a craving appetite.

We find that radical changes have taken place under the direct influence of the alcohol, which fully accounts for the dipsomania. The circulation in the abdominal viscera has changed, owing to the pathological condition of the liver, where we find a diffused inflammation of the connective tissues between the lobules. This inflammation produces an abnormal accumulation of amorphous granular matter, surrounding the hepatic lobules, and filling the inter-lobular space, causing pressure to be directed against the lobules, inducing severe inflammation, and finally atrophy and contraction of this important organ. This produces the condition sometimes called a hob-nail liver, because the liver presents in its atrophied state a nodulated appearance. Adhesions to surrounding parts often occur; as the morbid deposits are multiplied, the terminal branches of the portal veins are pressed upon, obstructing the flow of blood through the liver. Portal congestion follows, forcing the blood to seek new and unnatural avenues of escape; nature is not always capable of making this change quickly enough to avoid the effects of active congestion, and dropsy, or a hemorrhage from the mucous membranes of the stomach and intestines, ensues.

Compensating avenues of escape for the blood are, through anastomosing branches, shown by Sappey " to pass to the liver between the folds of the falciform ligament, and the ligamentum teres, communicating with the veins of the abdominal parietes through the epigastric, and internal mamary vein."

Other important avenues of a collateral circulation, are through the inferior mesentric and hypogastric veins, belonging to the hemorrhoidal plexus ; also by branches formed between the vena-portæ, and the veins situated in the serous covering of the liver, and newly formed vessels created in the tissue making up the adhesions of the liver to the diaphragm.

The superficial abdominal veins are often enormously enlarged, assisting the physician in estimating the extent of the lesion.   Jaundice is not a necessary concomitant, although in some cases it results from pressure made upon the minute biliary ducts.

While the liver presents marked changes, the power of alcohol to pervert and disorganize is not confined to that organ ; all the functions are disturbed.

Alimentation and assimilation, constituting the fundamental processes essential to life, are perverted by the alcohol, being suppressed in some respects and exaggerated in others, with but few compensating or conservative effects to modify these results. The ability of the absorbents to appropriate a normal quantity of nutritive principles from the food as it passes through the alimentry tract, is greatly impaired.

The system requiring support, demands the alcohol, declaring its want through a craving appetite for that which it can assimilate in its unnatural state.

The molecular deposits equalizing and maintaining a balance between the appropriation of nutritive material, and the destructive tissue waste, which is taking place in the economy continuously, ultimately change the composition of the tissues; the deposits intended to support the system by replacing the waste of disassimilation are not sufficient to keep the system up to a normal standard.

In a large majority of cases this condition is rapidly produced, as only a small percentage of the alcohol is discharged from the system. The victims carry an enormous amount of this amorphous matter, simulating normal fat, but consisting of a substance that undergoes a metamorphosis after being deposited. Owing to peculiar effects of the alcohol, the absorbents take up only an infinitesimal amount of solid aliment. In a few years the tissues are composed of this unstable mass, created by the liquor, and the functions of animal life progressively retrograde, as a hyperplasia of this adipose follows, and it is not capable of successfully resisting influences that would not affect normal tissue.

The capillary circulation is interrupted and slow, owing

to a lack of contractile force in the vessels, rendering them susceptible to the action of cold; tactile sensibility is blunted ; and insensible cutaneous exhalations are imperfectly performed, causing effete matters to be imprisoned, imparting a dark hue to the parts. The blood in peripheral vessels is imperfectly arterialized, and this gives the face and hands a purplish, puffy appearance.

The brain shares the general decadence; its capillary vessels are hyperæmic, and by a slow process softening in superinduced in many cases, producing that state of chronic imbecility so often observed in drunkards. Impaired sight and hearing are numbered among the complications; the kidneys, heart, and spleen receive, according to their ability for retaining it, fatty deposits that interfere with their respective functions and imperil life.

The normal activity of the kidneys depends upon a free and unimpaired secreting surface, which enables them to elminate freely the results of disassimilation, induced by muscular effort; when they become fatty and congested, dropsy supervenes. The muscular walls of the heart are rendered atonic from such deposits, and its contractions are apt to become paroxysmal in character when the system is excited by any unusual circumstance; this liability exposes the victims of alcohol to sudden death through a failure of normal contraction which causes an extravasation of blood in the brain, generally from the middle cerebral artery.

The degradation of character which accompanies

dipsomania is remarkable. The physical changes wrought by excessive indulgence in alcohol are not more extensive than the mental and moral changes. There is a complete suppression of the finer sensibilities, and an abnormal development of the grosser dispositions and desires, and this independently of the immediate existence of alcoholic stimulation. Intellectual dullness, weakness of memory, and timidity, are common accompaniments of the disease. The craving for alcohol becomes intense, and unless the disease is speedily arrested the end is near and certain.

The therapeutical indications are at once suggested by the pathological conditions presented. In aggravated cases, time, tact and perseverance are requisite successfully to combat the demon alcohol, and revolutionize the system, by controlling the dipsomania, and promoting assimilation.

While evidences of cirrhosis of the liver will remain for an indefinite period, the patient can nevertheless be fully restored to health and happiness. When the appetite for liquor is destroyed, the functions of digestion, assimilation and tissue metamorphosis become active; the finer sensibilities return, and the rescued victims can scarcely believe that they have been gutter-drunkards.

The conditions presented in dipsomania demand that judgment be exercised in estimating the capabilities of the patient's system to repair damage, and the degree of recuperative energy still in reserve.

If the patient is a stranger, the physician should find

by questioning, what his normal condition was before forming the habit. He will then be able to discover and locate the changes wrought by the liquor. If the patient's desires lead him to great excess, it is necessary to obtain at once decided effects with the remedies used. Individual peculiarities render it necessary to modify the treatment as circumstances demand.

The physician, whose faculties of discrimination are cultivated by careful study and observation, will take advantage of temperamental idiosyncrasies, for the relief of the patient, and the success of treatment, where a careless observer would meet with complete failure.

The presence of congestion, and the accumulation of effete matters, make necessary a preparatory treatment to excite cutaneous activity, and to stimulate the secretions generally, in order to eliminate naturally and vicariously all possible matters which endanger the system by their presence.

The patient should remain at home for a few days, beginning the treatment in the evening; as special advantages are thus secured. The following preparation should be administered at bed-time:

R Calomel, gr. x.
Podophyllin, gr. $\frac{1}{4}$
Elaterian, gr. $\frac{1}{16}$.
Capsicum, gr. i.

M. The calomel contained in the prescription should be triturated with three grains of sugar, to one grain of calomel, giving ten grains as combined.

The patient derives an increased constitutional effect from a small dose approximating the action of a large dose. We should assist peristaltic action of the bowels with a large warm water enema.

The patient can retire after taking a warm bath. It is well to paint the skin over the spinal column and right hypochondriac region with Tr. iodine, which gives tone to the nerve centres.

Jaborandi, in teaspoonful doses, should be administered every twenty minutes, until three doses are taken, to stimulate cutaneous transpiration.

If the bowels have not acted freely by morning, three teaspoonfuls of sulphate magnesia should be taken, or an enema of oil, as the possibility of the withdrawal of all alcoholic stimulants depends upon the complete relaxation of the bowels. If they have so acted by the following morning (and the physician should call early, not only to estimate and thoroughly understand the situation, but also to inspire hope and confidence by his presence, as the agony of reaction for the want of the usual stimulant will be upon the patient), the physician may administer, with full confidence in its capacity to sustain the system and ward off nervous irritability, and also to allay the desire for liquor, the following prescription:—

R.   Cinchona rubra, fl. ext. ℥ v.
                 Cannabis indica, ℥ iii.
                 Capsicum, Tr. ʒ iii.
                 Ginger, Tr. ℥ ii.

M. Sig. One teaspoonful every hour, decreasing the amount as circumstances admit.

If the patient's system is loaded with abnormal tissue produced by the alcohol, giving him a congested look, the taking of eight or ten ounces of blood from the arm will be followed immediately by favorable results. Ataxic symptoms denoting the approach of convulsions, epileptiform in character, are liable to present themselves at this stage, and should warn the physician not to hesitate, but boldly to abstract twelve to fifteen ounces of blood at once; the good effects thus obtained are beyond dispute.

If paroxysms are produced by the sudden withdrawal of liquor, superinducing hyperæmia of the brain, jaborandi and the bran-pack facilitate cutaneous transpiration, and allay symptoms of undue irritability. By persevering in the above treatment the appetite will be controlled, and the liability to delirium need excite no apprehension.

Premonitions of that state, however, consist in a hot, dry skin, torpid bowels, and congestion.

Under treatment patients usually pass the first day in comparative comfort, taking some nourishment, which should be offered but not forced upon them. Raw eggs, soups, and, if in season, lettuce dressed with vinegar, will prove acceptable, giving tone to the stomach.

If the alimentary canal has been thoroughly relieved by from seven to ten actions, ten grains of dover's powders can be administered at bed-time after the bath. Nitre paper will induce sleep if there is a tendency to insomnia.

During the next five days, from four to six doses of the prescription will prove sufficient to sustain the system. There will be no further necessity for interference upon the physician's part, unless the bowels become torpid, and active febrile symptoms appear, when the diaphoretic action of the packs and jaborandi will be called for.

On returning to business the patient should take a small bottle of the medicine with him, so as to follow instructions to the letter, as regularity in taking the dose is highly important.

Regular meals, served on time, should be provided, as a lunch of an inviting character will often save one from indulging in a drink, as hunger stimulates a craving for liquor, suggesting a desire for it, and food thoroughly allays this craving.

Keeping constantly employed is a safeguard that should not be lost sight of, and friends can help by providing occupation for the newly reformed, who usually have lost their business engagements during their dark night of unbridled indulgence. It is a powerful incentive to remain sober and reclaim lost manhood, for them to feel that a good start has been made, and that they are not left without the right or opportunity to help or advance themselves.

The medicine may be reduced during the second week,

to three teaspoonsful per day, taken before meals. If insomnia is complained of, the remedies suggested for that condition in the crisis of opium may be employed, and under no circumstances should we administer chloral, bromide of potassim, or opium.

184 TIlE OPIUM IIABIT AND ALCOHOLISM.

CHAPTER XXVIII.

Delirium tremens is caused by prolonged excesses in
the use of alcoholic drinks, and is finally excited by a
toxic element in the blood, accumulated through an
inability upon the part of the kidneys to eliminate excrementitious waste, involving congestion of the liver,
and partial paralysis of the nerves presiding over cutaneous activity. The brain receives the septic influence.
The intensity and peculiarities of delirium enable us to
estimate the amount of poison retained in the system,
and indicate the extent and character of treatment necessary for the victim's relief.

It is a difficult task to select a typical case, as delirium is presented under a variety of circumstances. We
call to mind a patient who gave us trouble and proved
for a time incorrigible, but who had never been known
to be intoxicated. He had drank liquor and beer every
day for twenty years prior to the attack, without exciting
ing unfavorable symptoms. The accumulation of effete
matter was slowly induced, and the equilibrium of the
nervous system gradually destroyed.

Sometimes delirium will quickly follow a prolonged
debauch. The victim during his debauch will search
out the lowest liquor shops to be found, and there drink

diabolical liquor, consuming little if any nourishment in the form of solid food. The bowels become torpid, superinducing hepatic congestion. The nerve centres imbibe by continuity toxic elements contained in the blood, causing delirium of a severe type to follow.

When the deposits which induce delirium are slowly accumulated, premonitions of its approach are observed. Birds of beautiful plumage are seen by the victim; monkeys performing antics, and other strange yet pleasing sights are present to the diseased imagination. If the blood is not relieved of this *toxical influence by appropriate treatment, symptoms of a different character will present themselves, indicative of a progressive change which places the patient in a precarious situation. He is frightened by unusual and imaginary noises; red spots in the carpet are mistaken for fire; the victim attempts to escape from some unseen foe. In the advanced stage of the malady, the victims of delirium experience an agony of fear by having snakes coil about, imagining themselves environed with hissing serpents, and other loathsome creatures; and they not infrequently commit involuntary suicide in their efforts to escape from these horrors.

Delirium is present in either condition. and the same exciting cause exists in both classes of cases. The patient who rushes into his parlors and dashes bucketsful of water over red spots in the carpet which he mistakes for a fire, at the same time giving his wife the suggestive advice not to get excited, presents a certain phase of the malady consistent with the amount of toxic matter in the blood.

In aggravated cases, the victims of delirium are liable to sudden death from paralysis of the heart. The immediate necessity in all stages of the disease, is to excite the dormant secretions to action and eliminate accumulated poison as expeditiously as circumstances will admit, thus removing the exciting cause of the malady.

The question is asked, why some persons of peculiar temperaments have delirium when they stop the use of liquor, and only then. While drinking regularly the abnormal standard created by the use of liquor is maintained ; when the liquor is withdrawn, relaxation, with irregular circulation, ensues, nature makes spasmodic efforts to throw off the poison, which prove unequal to the emergency. The adipose tissue is surcharged with that which soon becomes septic matter when poured into the circulation in the form of products of disassimilation, under the influence of relaxation. This unstable mass is rapidly transformed when unsupported by alcoholic stimulants.

The kidneys, disabled to a certain extent by fatty deposits, are not capable of accomplishing the task of separating and eliminating the excrementitious matter so abundantly produced. The result is only too apparent ; the effete matter being retained in the circulation, its poisoning effects manifest themselves in delirium. If the system is now placed under the influence of alcohol, the functions are again unnaturally stimulated to increased activity ; tissue change is arrested, and the delirium allayed. With the pathological conditions presented, it is difficult to believe that opium has been recommended during the first stage. The mortality where it has been resorted to, is a fitting rebuke to those whose therapeutical knowledge is not based upon a sound pathology. The use of opium in the first stage of delirium tremens is positively counter-indicated, as that drug closes the secretions, imprisoning toxic elements within the system, and increasing the accumulation of solid effete matters in the blood.

In placing a patient under treatment, the same methods that have been suggested in the preparatory management of dipsomania can be depended upon to excite free elimination. After administering the pill composed of calomel, podophyllin, elaterian and capsicum, eight to ten ounces of blood may be taken from the arm at once, after which, the hot bran-pack should be used, with ice on the head. If venesection and packs do not control the delirium—and in aggravated cases these measures often fail fully to allay paroxysmal effort—until after the bowels have moved freely, small doses of brandy compounded in egg with capsicum, may be given. During the interval, the physician should direct all his energies to the work of exciting the bowels to action with large enemas of soap and water, followed by oil to be retained.

When called in consultation where patients had suffered active delirium for three or more days, and the bowels were sealed almost to paralysis with opium or what not, we have resorted to single drop doses of croton oil with good effect, and have also used the battery for the same purpose with satisfactory results.

The importance of clearing the alimentary tract by relaxing the bowels is very great, as it facilitates the removal of hepatic and general congestion, and carries off an immense amount of effete matter vicariously. The delirium will abate and fever subside. Jaborandi may now be given, as hitherto suggested, and all liquor stopped. The following preparation can be administered every hour :

> ℞.   Cannabis indica, fl. ext. ℥ ii.
> Cinchona rubra, fl. ext. ℥ iii.
> Ginger, Tr. ℥ ii.
> Capsicum, Tr. ℨ ii.

M. Sig.   One teaspoonful in water as directed. After the secretions have fully responded to these remedies with several passages from the bowels, dover's powders may be given for their anti-spasmodic and soporific effect, with an assurance of favorable results. The appetite will also respond to the combined action of cannabis indica and dover's powders. Before offering food, however, it is well to allow the patient to rinse the mouth with vinegar and water, as a coated tongue or sordes on the teeth seriously interfere with an uncertain appetite.

Nourishment may be given by enema, in the form of concentrated soups, if the stomach will not retain it. Minute doses of chloroform will control vomiting caused by local irritation, enabling the patient to retain food. After the delirium has subsided, the baths should be continued, and the patient treated for alcoholism by the methods already suggested for dipsomania.

Persons suffering with delirium believe implicitly in the reality of their visions of snakes, monkeys, etc., regarding the testimony of their senses as the best of evidence. They cannot take into consideration their own diseased state, which creates a distorted and exaggerated vision. If the physician informs a delirious patient that no snakes are in his bed, when the victim not only sees but feels them, he commits a grave mistake, forfeiting the patient's confidence, and often creating the impression that the physician is conspiring against the patient, whose illusions he combats. It is cruel to hold or tie a patient; force enough should be present to guard windows and doors; the patient should have all possible liberty, and every appearance of antagonism should be avoided.

We were once hastily called at night to a case of delirium tremens. The victim's fearful cries and ravings could be heard for a distance of two blocks. We found that he had been "seeing snakes" for three hours past, and his friends, becoming frightened at his increased agony of fear, had securely tied him to a bed. We directed the attendants to cut him loose at once, and a look of satisfaction and happiness lighted up his face as he stopped his cries. In answer to our questions, he said large black snakes were coiling themselves upon his breast, while small ones were attempting to get into his mouth. We assured him that they had no fangs, and that he could surely remain where we willingly stayed. His distorted reason was thus successfully appealed to, and he gained confidence in our ability to protect him.

The acute symptoms responding quickly to the abstraction of fifteen ounces of blood, and a warm bath, the man was free from delirium the following morning.

Persons who have had delirium tremens are profoundly affected by the recollection of the agony of fear and horror through which they have passed, and they have a lively dread of a second attack. It is well to inform them that each subsequent attack will be marked by increased suffering, and that renewed attacks are likely to follow further indulgence in liquor. The circumstances render the time opportune for placing such patients under treatment for the diseased craving, with assurance of final success.

The nerves presiding over the diaphragm often become irritable, causing violent hiccough, which calls for interference. It can be immediately stopped by having the patient hold his breath and at the same time put a finger tip in each ear, while an assistant, acting in conjunction with them, closes the orifices of the nose—pressing at the same time on the inner canthus of the eyes over the lachrymal and nasal apparatus. This should be continued for a half or three-quarters of a minute. If the treatment should fail to control spasmodic effort, ice should be placed suddenly over the region of the ensiform cartilage, or this region should be painted with iodine. Chloroform in fifteen drop doses, as a final resort, seldom fails to give relief.

When treating the liquor habit in any of its varied phases it is a good rule, applicable also in the treatment of all bad habits, to have the patient attend as far as possible to his regular business during the time.

In nervous complications it is important to control the mind, as its influence is exerted for or against the treatment. It is, therefore, necessary to keep such patients busy, and their minds engrossed with details of business, allowing them no time in which to analyze their feelings.

This is without doubt the principal reason why inebriate asylums and sanitariums have so generally failed in their endeavors to accomplish permanent results of a favorable character. After remaining a certain period in the asylum the patients are obliged to face the world with all its temptations. Their return to old surroundings is so suddenly accomplished that the desire to resume their broken but still familiar habits is very great. If a radical cure cannot be effected while the patient is subjected to his usual temptations and at home, his friends must accept the melancholy fact that his chances for reform are not likely to be improved by a sojourn in an asylum. In our experience with this class of cases we can call to mind but one instance when a permanent cure was established in an asylum.

The sanitariums of to-day have degenerated from their original purpose, being transformed into temporary abodes for the convenience of patients who are passing through the crisis following a prolonged debauch, or recuperating from delirium tremens, and who are in no condition to be tolerated at home. They leave the asylum at the expiration of a week or two, without receiving treatment directed toward a permanent cure.

The effect of these periodical calls, made by this class of chronic inebriates, is pernicious and demoralizing on the few permanent patients who are trying to reform. They will smuggle liquor into the asylum in spite of watchful efforts to prohibit it, and regular patients who have not drank for weeks will be betrayed. Habitual drunkards are catholic in their tastes for companionship, and affiliate readily with any one who, like themselves, is addicted to liquor, managing to get on terms of intimacy within an hour after meeting. They are also easily influenced, and the persons they are necessarily thrown in contact with at an asylum, are just as bad if not worse than their companions at home.

The novelty of new acquaintanceship adds greatly to the influence of the occasional inmates over them. Fifteen or twenty idle men of this class would perish from inanity if they did not endeavor to outwit the asylum manager.

We should have a state law compelling inebriates to abide by their contracts, voluntarily entered into with asylums when they apply for treatment, with respect to time and other necessary restrictions. Favorable results would without doubt follow.

When a patient is suffering from the effects of a prolonged debauch, and is enduring the agony of remorse—with his nervous system shattered and unstrung, we should not allow him to be approached by persons wishing to give advice or to upbraid the victim. Religious precepts forced upon such persons, when racked with nervous irritability, are ill-timed and out of place, and tend to destroy the physician's influence over them. During the acute stage it is necessary to be considerate and to administer to the patient's wants with lavish gentleness. Such kindness is a more effective reproach than bitter upbraidings. When the patient is fully recuperated, timely advice and religious counsel may be given, but not before.

Considerate attention appeals to the better nature of the unfortunate victims of drink, while continual reproach excites them to rebel. They are naturally emotional and easily affected, and if appealed to through continued kindness, a large number of them will endeavor to heed the advice of friends. When they succeed in mastering the habit, we should not hazard an attempt to accomplish too much. We should not urge them to stop the use of tobacco—if they are addicted to it—until the system has adapted itself to changes already made. Such persons are often anxious to break all bad habits while their good resolutions are strong. It is well in such cases to restrain any tendency to overdo the matter, as reaction is apt to follow, imperiling the advantage already gained. Over-confidence in their own ability successfully to resist temptation is liable to cause a fall, and friends should earnestly warn them to be constantly on guard against the wiles of their old enemy.

CHAPTER XXIX.

.

NATURAL TENDENCIES.

WE do not care to speculate upon the question whether or not an inherited tendency exists with many persons, predisposing them to strong drink. Our experience suggests that a constitutional condition exists with some persons which predisposes them to this class of habits, making it necessary for them to exercise their will-power and to keep a continual watch upon themselves in order to curb all desire for liquor or the narcotics. It is not an essential prerequisite to the existence of this constitutional tendency that parents should have liked, or indulged in, liquor or tobacco. We have observed a number of cases among men and women from the higher walks of life in whom a tendency of this sort was manifest, although their parents, in many instances, were strangers to the taste of liquor.

On the other hand, who has not known drunkards' sons, left early to govern themselves in their own way, who have grown up to be sober men, respected for their many manly virtues. A fact familiar to all careful observers, is that two boys, neighbors from youth, subject to the same surroundings, enjoying equal religious advantages, will go out into the world and pursue

opposite courses.   One can successfully resist temptation
without an effort.   He is well contained, and exercises
perfect control over his appetite.   His companion,
possessing equal natural ability, cannot resist temptation,
and his family soon bewail his early fall.   The difference
is in temperament, not in training.   Parents with young
boys to train, should bear this in mind.   Boys possessing
superior mental endowments, are capable of accomplishing
great good, in a life full of lofty purposes; they can also
go to fearful excesses and sink low in the scale of human
degradation.   Few parents anticipate the coming storm,
until it is upon them in all its relentless fury.   I ad-
monish them to be watchful, and not to place too implicit
reliance upon the fact that they have not themselves
transmitted a craving for liquor to their children.

There are so many influences brought to bear on the
young of to-day, to induce them to drink, and pathological
changes are so rapidly produced by indulgence, owing to
the character of the liquor sold, that safety is found only in
total abstinence.   This should be secured by destroying
the desire for strong drink, and the ability to consume it.

To engender a permanent distaste for intoxicating
liquors, a systematic treatment, conscientiously carried
out, is necessary.   It requires, upon the part of attendants,
an unflinching determination to follow out to the letter
all instructions.   As a necessary precaution, the immediate family should be informed that the treatment
is liable, owing to its severity, to impress those who are

not familiar with its course and results, as perilous to the patient's after-health, such is not, however, the case. The method appears cruel in the extreme, and the persons treated do suffer to some extent. They may be subjected to treatment as early in life as circumstances will admit, the young being susceptible to more perfect control than is possible with older persons, and impressions received are retained with an exaggerated vividness peculiar to youth. Any period subsequent to the fourteenth year, or as near that age as possible, will assure favorable results. The physician is, however, justified in treating boys previous to that age, if they become intoxicated, showing by that act their willingness to indulge as occasion offers.

The treatment is indicated for fast young men, whose natural proclivities are such as to desire liquor, as a social pledge, as they will not subject themselves to the cinchona rubra mixture, and accept its results by remaining cured. If they do consent, no permanent good will be obtained, as they generally take your prescription to please their family. The patient must remain at home under the constant observation of a trusty attendant. The liquors used in treatment including all that are usually sold, may be compounded together as suggested below. To destroy a patient's taste or desire for whiskey only would be useless. He could still drink champagne or beer until his sensibilities became blunted, when his ability to consume even whiskey would be restored, and in that case no good would have resulted from treatment. By using the following prescription patients are effectually debarred from ever consuming as a beverage any of the popular drinks that intoxicate. By seasoning all foods with the compound we impart a decided flavor of each constituent, sickening them of all.

R̥   Whiskey, ℥ ii.
            Sherry wine, ℥ i.
            Port wine, ℥ i.
            Lager beer, ℥ i.
            Gin, ℨ ii.
            Cider, ℥ i.
            Rum, ℨ i.
            Champagne, ℥ i.

Every particle of food must be thoroughly impregnated with this mixture before it is consumed. In fact everything taken into the stomach, except drinking water, must be flavored.

Meals should be served as usual, consisting of a variety, in order that the patient may not become forever disgusted with any one article of food essential to future alimentation. One half-teaspoonful of the compound is sufficient to flavor tea, coffee, or milk, the same quantity for meats, potatoes and bread. Side dishes and dessert must be treated in the same manner, not allowing a mouthful of anything to escape. The quantity is so small that no stimulating effect is experienced.

The patient often regards the treatment as a huge joke, during its early stages. Some have sarcastically informed us, that it would be wise to stop, and save the prodigious waste of liquor, hinting that the flavor was quite agreeable. During the second week the stomach revolts, and regurgitates the seasoned aliment, requiring the pangs of hunger, coupled with a liberal exercise of will-power, to enable the patient to consume even a small amount of the prepared food. The time for nausea or vomiting to appear is uncertain, depending, as it does, upon temperamental differences. Patients of lymphatic tendencies, being susceptible to extrinsic impressions, respond to the nauseating effects of the liquor within a few days, while persons of an opposite temperament will resist its influence for two or more weeks.

The time for stopping treatment must depend upon the above named symptoms. A rule applicable to the greater number of cases, requiring to be modified if circumstances demand, is strictly to confine patients to seasoned food, for a week or ten days after nausea presents itself. After nausea is experienced they will not desire meals at the appointed time, putting off the trying ordeal as long as possible. The attendants should allow them to suit their own fancy in the matter of time, preparing food as it is called for. They are suffering the pangs of hunger, rendering the desire for food not seasoned with liquor very strong, and a strict watch must be kept upon them to guard against the possibility of their breaking treatment, by taking food unprepared.

Their gnawing hunger tempts them to eat, and they

endeavor to accomplish the act, only to be sickened by the smell and taste of liquors. They are becoming disgusted with the very name of alcohol. Although ravenously hungry they cannot control that disgust for the liquors sufficiently to take food flavored with the mixture. Parents are liable to become unduly solicitous, fearing that injury may be sustained from starvation, as the patient generally declares with tears that he is starving It is well to remind them of the physiological fact, that a human being can go for from fifteen to twenty-five days without food of any kind.

Patients will be impelled at last by extreme hunger to bolt a few mouthfuls of food, only to throw it off. They try repeatedly, with the same result, until they so abhor the taste and smell of liquors, as to resign themselves to the horrors of starvation rather than endure the taste. They will go from twenty-seven to thirty hours without aliment of any kind. They should be allowed to go that time, although weak and prostrated, appearing as if they had sustained a long attack of sickness. The physician must be the judge when to suspend treatment, and he must resist the tendency to stop too soon, influenced by sympathy against his judgment.

After a certain stage has been reached further treatment would be a needless infliction of pain. If an error is however to be committed, let it be on the side of continuing too long. A few days' unnecessary suffering will only result in the pain endured, not inducing permanent damage, and effectually curing the patient. On

the other hand, if the system is not allowed to come
thoroughly under the influence of the treatment so as to
excite lasting disgust, the result will be similar to that
which young boys experience when forming the habit of
smoking or chewing tobacco, they stop when sensations
of nausea are suffered, and wait for a time when they
try again with better success. Occasionally they will,
through fear of being laughed at by companions,
persevere and resist the nausea until thoroughly under
the influence of tobacco, when they are exceedingly
sick. The effects are salutary, however, as the excess
forever destroys their ability to repeat the act, and even
in old age, their disgust and horror of tobacco will be
acute. Patients treated as recommended eight years
ago, declare their nauseated disgust for liquors to be as
acute as it was immediately after treatment, and they
avoid passing near liquor-shop doors, through fear of inhaling the sickening fumes. When the patient resumes
his customary diet, the same precautions should be observed that would be necessary if the condition was presented under any other circumstances. The only complication observed, is a sympathetic reflex irritability of
the stomach during the latter part of treatment, causing
obstinate vomiting to be excited by the sight of food
similar in kind to that which the patient has tasted with
liquor in it. The physician must have him shut his eyes
on returning to regular diet until his functions become
normal. Patients recuperate rapidly, regaining lost
tissue within an incredibly short space of time, and

entertain during life an undisguised hatred and disgust for the smell and taste of intoxicating liquors

---

CHAPTER XXX.

CHLORAL HYDRATE.

WHILE not possessing the active power to produce evil results that bromide of potassim has when habitu-ally used, chloral hydrate is in a peculiar manner treach-erous and uncertain in its action, demanding close atten-tion upon the part of physicians in its administration. As with other powerful hypnotics, it is often resorted to in procuring that priceless boon—denied to so many—an unbroken night's sleep. Owing to its charming and apparently harmless effects, temptation to resort to it is beyond the ability of many persons to withstand.

When consulted by patients suffering insomnia, a grave responsibility is assumed by the physician who directs that thirty grains of chloral be taken on retir-ing—or who writes a prescription for the same, as patients renew their supply of the drug when they think that the symptoms justify them in doing so. They do this innocently, or to save the cost of procuring a second prescription. A great danger attends the administra-tion of this drug, because no time supervenes in which restoratives can be applied, after alarming symptoms present themselves, as death follows quickly, rendering the physician's efforts unavailing. Just what the conditions are that lead to this result is a matter that has never been fully determined. The phenomena denote failure of the heart's action. A robust lady, in the prime of life, suffering with a decayed tooth, died in twenty minutes from the effects of ten grains of chloral. We call to mind the case of a gentleman who had been laughing incessantly for seven hours, when he came under our observation, and who was quite exhausted. He had taken twenty grains of chloral, eight hours before, this being one of its peculiar effects. Many physicians suppose the drug to be cumulative, as in some cases no appreciable effect has followed its administration every half-hour in doses of ten grains, and yet the fifth dose has caused sudden death, by asphyxia.

The habitual use of chloral slowly but surely disorganizes the red corpuscles of the blood, impairing their integrity as carriers of oxygen, and so undermining the economy, weakening its power of resistance against outward impressions. The same dose that was taken in safety in beginning the habit, eventually proves fatal. The necessity for exercising caution forces itself upon us. It is best to prescribe chloral only when positive indications exist for its necessity. In mania-a-potu, seventy grains of chloral have been given with no unfavorable results. It is usually taken, however, as a means of procuring sleep, and a habit is formed, not in obedience to an imperative demand to ward off relaxation, but because after resorting to it for any length of time the ability to rest without a certain dose is destroyed.

The exciting cause is not a congested state of the brain, but a peculiar activity of the mental faculties, which precludes sleep, and is only allayed by chloral. The drug acts by contracting the minute vessels of the brain, causing a profound sleep—by withholding the amount of blood necessary to normal activity. The practice induces in time poverty of the parts involved. If the victims could sleep, the nervous system would not be painfully excited, as it is with opium or liquor under similar circumstances. Owing to the peculiar action of the drug, insanity will follow insomnia for the want of chloral, making its appearance before any premonitions of a nervous character are observed.

The drug cannot be withdrawn gradually, as any amount under the customary dose will give no results whatever. The necessity for stopping its use as a habit, to save life and ward off insanity, is imperative. The progressive poverty of the nerve centres following the continued indulgence accounts for sudden death—the fate of so many. While opium interferes with tissue changes, it nevertheless stimulates the functions of life in other respects. Liquors also compensate to a limited extent for the damage done by them. But chloral robs the system—exciting sleep by depressing functions essential to mental activity.

The indications are for the immediate withdrawal of the drug, and the use of means to induce rest for a time until a normal degree of vitality returns to parts impoverished by the action of chloral, after which natural sleep will be enjoyed. Owing to the fact that it is impossible gradually to reduce the drug, vigorous methods of treatment must be adopted. If painful complications are presented, they should be managed as if recurring under any other circumstances. In connection with soporifics, supportive and tonic remedies are necessary to increase the red corpuscles of the blood. If these means fail to produce sleep, and insomnia is suffered for two or three nights, hysterical symptoms will present themselves. The patient hears strange noises, the prolabia is white and bloodless ; anorexia is complete, while the eyes are unnaturally bright, the conjunctile membranes become puffy and congested, suggesting kidney complications ; more or less of gastric disturbance is present while nature is making an effort to adapt itself to the change. The abdomen is distended, giving a tympanitic resonance, denoting the presence of gas, and for that reason food should not be forced upon the patient, as digestion is imperfectly performed and there is a decomposition of aliment. Persons who use chloral usually take but one dose in twenty-four hours—generally at bed-time—consisting of from ten to forty grains.

During treatment favorable hygienic surroundings should be secured, and the patient should have a light, airy chamber, with bathing facilities. In treating the results of chloral a liberal supply of money is necessary

to secure success and comfort, as one of the principal
drugs is very expensive, viz., pure Chinese musk. An
impure or adulterated article utterly fails to give any
results whatever; pure musk, costing from twelve to
fifteen cents per grain, must be used. It is the only
drug known to us that effectually corrects the peculiar
state following the chloral habit. The office performed
by cinchona rubra in the liquor habit, and by cannabis
indica in allaying irritability excited by opium, is performed by musk in relieving the chloral habit. From
ten to fifteen grains may be administered at bed-time.
To facilitate its action baths, hop pillows, and nitre paper
can be resorted to, and cupping over the region of the
spine often gives good results. If progressive anæmia
preceded or accompanied the habit, the actual cautery.
used as has been suggested for pulmonary consumption,
materially assists in restoring activity of the nerve
centres, giving tone to their periphery.

After inducing sleep for a few nights, the musk can
be gradually withdrawn. A few mouthfuls of warm
soup, containing capsicum, taken at bed-time, will often
induce sleep by equalizing the circulation, and relieving
capillary congestion of the brain. A wet sponge fastened to the back of the neck and head, will also induce
sleep by acting on the base of the brain.

Dietary habits exert an influence for or against sleep,
and no substance liable to disturb quiet rest should be
eaten at night. Celery, lettuce and onions mildly favor
sleep. Milk can be relied upon, and it seldom gives

other than good results. Lime water and lacto-pepsin facilitate its action. Many persons declare their inability to use milk, but we consider this an affectation in a considerable number of cases. Milk contains as an aliment, a variety of nutritive principles essential to life, and without other food it has sustained all of us, at one time of life. It is easily assimilated, leaving but a small percentage of residue to be carried off. If its use should cause bilious symptoms, as some persons believe that it does, such effects will pass away within a few days, leaving only good results.

Free and regular actions from the bowels every day are conducive to natural sleep. The bowels should therefore be regulated as soon as possible, but not by the action of drastic purgatives, as no permanent good is attained in that way. Habitual constipation is maintained by two exciting causes, and both must be overcome; the first of them is a want of tone sufficient to excite peristaltic action. This is to be corrected through the nervous system. The second cause is the lack of a sufficient secretion to lubricate the mucous coats. If this secretion is induced by means of a drastic purgative, the secretive organs are stimulated to over-action, and a superfluous amount of fluid is poured out and carried off, impoverishing the over-excited functions. The materials discharged by secretion have to be replaced from the blood, and the patient has lost that which he can least afford to spare. The natural result is an absence of proper secretions for a few days, or until the wasted material is replaced by the system; and during that time constipation is more marked than it was before.

The indications are that the nervous system should be toned up, by the judicious use of phosphoric acid, strychnia, iodide of lime, ammonia iodide, hydrag-bin-iodide, ergot, iron, gentian, calisaya, quinia, or cannabis indica, careful consideration being given to the condition and needs of each case. Well chosen foods, known to excite a laxative influence upon the bowels, should be liberally consumed. With chronic cases dietary precautions will not be sufficient until a habit of body is established. In securing that condition the physician should avoid drastic remedies, as they will not only overdo, but create a state demanding their habitual use. The influence of the mind in exciting the nervous system to generate force at certain times, can be utilized and made to serve a good purpose, by insisting that the patient shall go to the stool every morning regularly after breakfast, whether a desire to defecate exists or not. The use of coffee should be positively interdicted, and to assist in forming habits of regularity, it is well to administer—

> ℞.   Magnesia sulphate, ℥ i.
>
> Senna, u. ext. ℥ iv.
>
> Aqua, ℥ x.

M. Sig. One tablespoonful at bed-time. We have accomplished the desired end in some cases by giving a tablespoonful of bran, mixed in a tumbler half full of water. at the same hour. With anæmic patients, oil in teaspoonful doses, taken every night, is preferable to this, and it may be indefinitely taken, as it not only acts upon the bowels, but is an aliment, and so the patient derives a double benefit from its use Brown bread may take the place of wheaten bread. Alteratives should be withdrawn as soon as a habit of regularity is assured.

If the oil is distasteful it may be taken by enema.

CHAPTER XXXI.

DYSPEPSIA.

Its treatment independent of and complicated with the chloral habit.

DYSPEPSIA is a troublesome disease, and is presented by a large proportion of patients addicted to chloral, not as a necessary result of the habit, but because when a tendency to dyspepsia pre-exists its development is rapid under the influence of the drug. As a concomitant to the habits of opium or chloral, it should be overcome if possible by judicious treatment, as in case of neglect, or a failure to accomplish a permanent cure, the indigestion remains as an exciting cause for a multiplicity of complications, predisposing the patient, through suffering, to form a new or continue old habits. As the necessity for a radical cure is of paramount importance, we give the subject careful consideration.

The term dyspepsia, as conventionally applied, gives us no intelligent explanation of the lesion sustained. All painful disorders of the stomach, occasioned by repletion or difficult digestion, and a number of acute and chronic conditions of different kinds, arising from a variety of causes, are designated dyspepsia. Morbid conditions should be fully recognized, and a differential diagnosis must be made before formulating a plan of

treatment. An uncertain method of forming hasty conclusions, and prescribing indiscriminately has caused
needless ill-health and suffering, driving patients to seek
relief at the hands of quacks, or to the use of nostrums.

Acute dyspepsia is easily excited with persons whose
peculiarities render them susceptible to it. Exposure to
cold deranging the secretions and leaving the mucous
coats in a state likely to inflame readily; the eating of
indigestible articles of food; and strong emotions arresting gastric activity, are common causes. Fatigue,
either mental or physical, will induce what many persons
are pleased to term a bilious attack, or gastric fever, the
latter symptom being purely symptomatic, and not es-
sentially a fever, as the name implies.

Sub-acute symptoms are a sense of fulness and pain
in the epigastrium, nausea and headache, accompanied
by general malaise. The tongue is coated, and there is
complete loss of appetite, with a sour, unpleasant taste in
the mouth. This condition is usually amenable to mild
evacuant remedies. Calomel in half-grain doses, triturated as before suggested with sugar, and a warm bath
to excite the cutaneous secretions to action, give good
results. The appetite generally regulates itself, and
food should not be forced upon the patient. Acute
gastritis is often presented to the practitioner, especially
in the spring, when the above described symptoms are
present, greatly aggravated. The bowels are constipated
in a large majority of cases; the tongue is heavily
coated; the breath is offensive, and there is a total loss

of appetite; vomiting accompanied by a severe headache, with febrile symptoms, is usually observed. A liquid containing bile is regurgitated; the urine is loaded with the lithates, and excites a scalding sensation in the urethra after micturition. The patient appears sallow and stupid, usually entertaining a dismal opinion of his condition, and having little hope of a final cure. The remedies suggested above, in large doses, adding podophyllin to the calomel, and facilitating their action by relieving the lower bowels with a large enema, and a warm bath or pack, will afford relief within a few hours. Tonics, including quinia, iron, and gentian, are indicated for a limited time after acute symptoms have been allayed.

While the sub-acute and acute forms are quickly excited, reaching their maximum of intensity within a short time, and are as rapidly overcome by well directed treatment, the chronic state presents morbid anatomical conditions, with structural lesions of the digestive organs, which have been slowly induced and which resist treatment, holding pertinaciously to the system. The symptom that first arouses suspicion upon the patient's part, declaring the presence of dyspepsia, is a painful or labored action of the stomach during digestion, which for a certain period remains a local complication. As pathological changes take place, symptoms of a sympathetic character, indicative of its general influence over the economy, are suffered. These symptoms show, the nervous system to be involved.

Digestion, instead of producing a sensation of comfort and satisfaction, as it does in health, excites feelings of uneasiness in the intestines, which seem to be full and bloated. The discomfort is often most extreme when the food appears to be thoroughly digested. Again dyspeptics sometimes suffer little, if any local pain, and yet digestion is imperfectly performed. The absorbents can not take up sufficient nutrition to nourish the body, and the patient becomes weak and anæmic, having at best a wretched existence. Regurgitation of food, both solid and liquid, is a symptom usually appearing early in the disease. Owing to putrefactive changes, the liquid is intensely sour, seeming to scald the throat. Again, it has a greasy, nauseous taste, derived from the presence in the liquid of hydrochloric, acetic, or lactic acid, from the stomach glands. Pyrosis, or in common parlance, water-brash, occurring usually in the morning from an empty stomach, is the result of a nervous condition, which affects the glands at the cardiac end of the stomach together with the esophageal glands, causing them to pour out an abnormal secretion. Cardialgia, called heart-burn, is a disagreeable sensation resulting from an excess of acid secretion. Tympanitis, following the indigestion of food, arises from a deranged state of the functions controlling digestion and absorptom.

The condition of the tongue near its base is indicative of the state of the mucous membranes lining the stomach. A low form of inflammation in the stomach causes an exudation to be thrown out that covers the membrane, deranging its secretions and impairing the activity of its absorbents. The stomach cannot therefore perform its functions normally, and food is impelled indigested into the intestines, where a partially successful effort is made to digest vicariously that which the stomach has allowed to pass, causing putrefactive and fermentive changes to take place, and generating enormous quantities of gas. If eggs are consumed, sulphuretted hydrogen is produced; other articles of aliment generate gases peculiar to themselves. The amount accumulated often interferes with respiration, and with the heart's action, creating a fear that the organ is diseased, when in fact its action is only impeded. The gas crowds the diaphragm into the space allotted to the heart, and painful action is sure to follow. Many cases are recorded, however, where the pressure was so great as to retard the heart's action, and stop peristaltic movement of the bowels, causing congestion of the liver and spleen to follow—death relieving the sufferers from the terrible agony induced within a short time.

As structural changes involving the mucous coats and absorbents become established, constitutional disturbances supervene, the persons affected lose flesh progressively. The countenance has a tired and haggard look, or such a look as it would wear if some impending danger were constantly apprehended, varied only by a sleepy, vacant expression of pain. Persons afflicted in this way cannot concentrate their thoughts upon any subject for any considerable length of time. They become peevish and habitually low-spirited. An irresistible desire for sleep will oppress them, at certain hours of the day, only to disappear at bed-time, rendering their night's rest unsatisfactory. They are disposed to settle into a helpless state of apathy, from which friends experience great trouble in arousing them. They are also apt to think that their cases are misunderstood by physicians, and that they are victims of some obscure disease, which is gradually carrying them to the grave. A slight cough occurring night and morning, excited by an adynæmic state of the nerves, keeps alive a suspicion that consumption has marked them. Their appetite in the morning is capricious, and they have at unseasonable hours an abnormal desire for some special article of food that is exceedingly sour, or very sweet. If this craving is indulged, the reaction is painful and prostrating.

They are constantly on the look-out for strange symptoms, endeavoring to discover new manifestations of disease. They count the beats of the radial pulse, or the number of respiratory acts, per minute; or they imagine that impotency or insanity is imminent. In this state of uncertainty they are liable to become addicted to the use of liquor or the narcotics, and so to destroy forever their last chance of restoration to health. Liquor excites peristaltic action, and quickens gastric effort for a limited period, leading the victims to consider it a specific remedy for their many troubles. The use of liquor places such persons in a very dangerous position, as it still further complicates the disease, by increasing the exciting structural lesions.

When the disease has been suffered for months, the cuticle indicates quite clearly the condition of the mucous coats covering the stomach, showing to what extent the nervous system has been impaired in its capacity as a generator of force to maintain circulation in the cutaneous capillaries. When the nervous system is thus incapacitated for furnishing a sufficiency of nutrition to the coats, the cutis will be cold, or unnaturally hot, thick in appearance, and full of pigmented deposits, described by many persons as moth patches, all of which should have been carried off, and eliminated. The coats lining the stomach, containing glands and absorbent structures, are in the same abnormal state, precluding the possibility of natural assimilation. We have noticed when constipation is present and the urine is loaded with crystals of oxalete of lime, that melancholia and hypochondriasis are apt to be developed.

There is a wide diversity of opinion as to the exciting cause for dyspepsia. It still remains a mooted question. Excess in eating is supposed by many persons to be the principal cause, but we are constrained to doubt this assumption. The largest eaters ever observed by us have been entirely free from dyspeptic symptoms. The manner of consuming food undoubtedly plays an important part as one of the exciting causes with those who are predisposed to the disease.

Young persons in the midst of their studies sow the

seeds of future dyspepsia by overdrawing on their nerve supply. The functions of animal life are naturally heavily drawn upon, in appropriating the material neces-sary for bodily growth. Beyond a certain point this constant draft to supply the brain with the requisite force to accomplish an imposed task is over-work. The mental faculties are cultivated, at the expense of some important physical function. The gastric functions are actively employed a large part of the time in replacing constant tissue waste, and for this they demand a liberal supply of nerve energy. If this energy is not forth-coming—because it has been already spent in support-ing undue mental exertion—an initial step towards an abnormal state is taken. Young students still further increase the danger by eating rapidly, and as rapidly turning to think of something else. Mastication is im-perfectly performed, and work is imposed upon the stomach for which it is physically unfit. The result is obvious. Food properly prepared by the teeth, passes into the intestines, after being emulsified in the stomach, while lumps and indigestible particles are repeatedly thrown back into the stomach, by contractions of the pylorus which prevents the exit of food not sufficiently emulsified.

Unprepared to perform labor of that character, con-tinually imposed upon it, the stomach rebels, and a low form of inflammation is induced, resulting in an exuda-tion—similar in character to that observed on the tongue, which impairs glandular activity, and interferes with the normal action of the absorbents. The next meal is not fully assimilated, and the system is obliged to accomplish labor for which it receives no compensation in nutrition.

The system begins to fail, the victim becoming weak, dull and languid, as the nervous forces for the time being are concentrated upon the work of repairing the damage. The breath is bad, the head heavy or aching, and the sufferer is cross and peevish. Parents are tempted to administer stimulating condiments, or pepsin, to excite digestion, while urging the afflicted children forward in their studies. By these means children are not infrequently condemned to a life of ill-health, predisposing them to acquire the dementing habits. Instead of this, all mental labor should be suspended at once, allowing the nerves time in which to equalize their forces, assisted by plain, nutritious food, slowly masticated, and the child should have an opportunity to take an abundance of fresh air, out-door exercise, and sunshine.

Business men approaching the prime of life, who have inherited from their progenitors well balanced systems, sometimes inadvertently undermine excellent constitutions by excesses. The hurried, feverish lives led by ambitious business men in the struggle to acquire wealth and position, keep the nerves strung to a high tension, with continual anxiety and mental disquietude. Such a strain not only sows the seed of dyspepsia, but is liable to be followed by other ill results. The hurried man of business takes his cup of coffee, containing but an infinitesimal amount of nutrition—besides the sugar and milk which it contains—and bolts an indigestible roll, after which he starts on the run for a car, and by the last act, he stops digestion, and robs his insulted stomach of the miserable portion provided for its susten ance. He takes luncheon at eleven, twelve, or two o'clock, as circumstances attending business allow, talking longs, shorts, corn, or what not during the time, con- suming but little solid food, but drinking large quantities of liquids. Returning home to dinner at six or seven o'clock, he gets the first real meal of the day. He now has the time and disposition to eat heartily, at a time when he should indulge lightly, and yet he considers it strange that sleep is not enjoyed. When such men do rest, it is in a deadening, heavy sleep, awakening from which they complain of being unrefreshed and tired, with a headache threatening. The appetite is precarious, only coffee being wanted in the morning. Digestion is difficult, and attacks of melancholia appear, accompanied by worry about business matters, that before gave no con- cern; there is also a tendency to look on the dark side of life. Food is regurgitated, constipation is suffered, and in exceptional cases diarrhœa is present. The af- flicted person grows peevish, loses flesh progressively, or presents a bloated, jaundiced appearance.

If medicines are taken, only temporary results follow, as the exciting cause is left untouched, and general ill-health is rapidly induced. Changes suddenly made, as from a sedentary life to one of activity, predispose such persons to dyspepsia. As the nervous system, by years of habit, has adapted itself to a quiet life and is not prepared to generate an extra amount of force upon a moment's notice, the gastric functions are the first to suffer from the change. A change of an opposite character will induce equally disastrous effects, as the system accustomed to out-door exercise, on assuming a quiet, inactive life, will not eliminate solid effete matters, and the blood will become loaded to an extent, causing inflammation of the mucous coats, before the system can adapt itself to the change.

The necessity of continuing to transact business which is distasteful and repulsive, causing continual dissatisfaction, provokes dyspepsia. It is not always that the patient who has dyspepsia suffers pain in the stomach or regurgitates food. A chronic paralysis of the absorbents sometimes exists, exciting a long train of evils, affecting its victim both mentally and physically, but produces little local pain. Mental strain of whatever character invites the disease in its most aggravated form. Epicures and gormands happily situated, seldom have gastric troubles, while we have known moderate, careful men, with scolding wives, to suffer severely from dyspepsia, and in many cases they have not responded to treatment until in despair friends have been obliged to hint that a change upon the wife's part would be con-ducive to the sufferer's health and longevity.

Mental disquietude, or a fear of some impending

calamity, robs the stomach of tone necessary to digestion. When the anatomical arrangements of the nerves
presiding over the gastric functions are examined, we
are not surprised that such is the case. These nerves
interlace with the spinal and cerebral nerves, and it is
only natural to infer, that mental depression will produce partial paralysis, by a reflex action. Those persons who have bright aims in life, and by well directed
efforts are gradually proceeding towards their realiza-
tion, seldom have the disease. The over anxious, who
are chafing to break the bonds that hold them, fall
early by the wayside, victims of dyspepsia.

The lower class of day laborers are not much subject
to it, as they have few wants that are not satisfied in a
reasonable measure, while the mental faculties are not
brought into play, as such persons are not stimulated by
ambition to acts requiring mental endurance and do not
experience the exhausting strain of hope deferred, and
envy. As a class common laborers are contented with
minds inactive. The greater number of victims belong
to the great mass of educated people, who only lack, as
they imagine, the one essential pre-requisite, money—to
have their claims to social position recognized. They
strive to acquire wealth, bending every energy to the
accomplishment of that end. Some few attain it, and
retire to enjoy their well earned position, only to awake
to the fact that real pleasure is to be found only in
successful effort. Sudden reaction caused by retiring
from active business pursuits excites dyspepsia with

its train of ills. Retired business men are idle, and correspondingly miserable. Others striving, fall by the way, and so, never know how little they have lost.

In treating the conditions embraced under the name dyspepsia, if we would secure for our patients permanent results, it is of paramount importance, that a correct appreciation shall be had of the extent of the lesion suffered; and that the exciting cause shall be thoroughly understood. Each individual case will present symptoms, if they are cautiously searched for, suggesting the therapeutical measures necessary to success. If the exciting cause is an inherited nervous tendency, and the physician prescribes for an inflammation, supposed to arise from dietary indiscretions, the failure will be as humiliating as complete. It is necessary to discover the exciting cause, and remove it, and nature under favorable conditions will work a cure of itself. Medicines are necessary, but should not be depended upon to the exclusion of hygienic measures and dietary regulations. Obstinate dyspepsia is provoked by an insufficiency of nutritious food. Many persons are influenced to adopt an insufficient diet by reading popular journals of health, in which the writers possessing more literary ability than physiological knowledge, frequently advise all men to adopt a spare diet, because such a diet happens to agree with the writers' own peculiar constitutions. The physician who ignores the existence of different physiological necessities—as represented by a variety of temperamental conditions, and attempts to formulate one uniform dietary rule for all men, will find that the results are unfavorable in about two-thirds of the whole number of cases.

In cases where the nervous system appears to be in a normal condition, and it is found that the dyspepsia is the result of habitual over-eating, continued for many months, combined with indulgence in alcoholic stimulants, a rest should be secured by permanently withdrawing the alcohol, and placing the patient on a milk diet for fifteen or eighteen days. He will suffer acutely the pangs of hunger for two or three days, and should be cautioned accordingly. The best of results have followed this treatment. Lime water mixed with the milk facilitates its digestion, and assimilation. On returning to a solid diet, such patients should be cautioned to masticate all food thoroughly, and to eat slowly in order to secure the stomach against the danger of repletion, by giving it an opportunity to declare itself satisfied. Farinaceous articles of food can enter into the daily bill of fare to a certain extent.

Dyspepsia accompanied by or following progressive anæmia, with persons who have been in ill-health, is due to a want of strength upon the part of the nerve centres to generate sufficient power, to keep up peripheral activity. Food is not appropriated, as the absorbents cannot perform their part in a normal manner. The physician must revolutionize the system, by imparting tone to the central organs. Before giving medicine under such circumstances, we use with good effect, the actual cautery over either side of the spinal processes, as a preparatory measure. If a scrofulous diathesis is suspected, the following prescription may be given to remove the original cause:

R   Iod. Pot, 3 v.
           Bi-chlo. hyd. gr. i.
           Iodine Tr. 3 v.
           Aqua, ℥ vii.

M. Sig. One teaspoonful in a wine-glass of water after meals. The bowels must be regulated before permanent good can be expected. The methods heretofore suggested may be resorted to in correcting their action. Regularity in taking meals at stated times is highly important, as the system is easily deranged by irregularities.

Special attention should be paid to details in arranging the table and its surroundings. The dining room should be light, airy and pleasant. If the dinner hour is prolonged into twilight or darkness, an abundance of artificial light should be supplied. Neatness and cleanliness of the table linen, with a display of artistic taste in arranging the table furniture, are very desirable. The family should respond at once when summoned to meals, and avoid needless detention that has a tendency to excite irritable nerves. They should also accustom themselves to appear happy and pleased. These matters, while insignificant in themselves if taken singly, are important as a whole in their influence for good. Table talk may be confined to topics of general interest, and its tone should be uniformly cheerful.

The diet should have a wide range, embracing a variety of nutritious articles, including an abundance of ripe fruit to be taken in moderation. Those afflicted with dyspepsia are especially apt to consider certain kinds of aliment deleterious in its effects upon them. This is a fallacious idea, and should not be encouraged when milk, rice, and other articles known by experience to be wholesome and nutritious, are rejected upon this plea.

When the appetite craves only coffee in the morning, the further use of coffee should be stopped. The patient will be agreeably surprised to find with what ease he can abstain after the fourth or fifth morning. During that period, however, he will suffer for lack of the stimulating effect of the coffee, and will complain of weakness, headache, loss of appetite, and a general feeling of dulness. The above described symptoms of relaxation passing off, the patient will relish his food, having a natural appetite for breakfast, and taking it with avidity. Becoming hungry, and not satisfying the demand at once, has a pernicious effect, as the sense of hunger excites to activity the glands concerned in digestion, and they pour their secretion into the empty stomach, deranging its normal condition by working chemical changes before food is brought in contact with the gastric juices.

When excessive hunger is experienced, four meals may be taken daily, as a precaution against the liability of over-eating at any one meal. When the system demands food between meals, the craving is not abnormal; it indicates that tissue waste is in excess of supply, and should be met by an increased number of meals. Tissue waste is constantly taking place, and we can as continually replace the products of disassimilation, by increased frequency in supplying nourishment. No more food need be consumed in the aggregate. The appetite thus dealt with will not be voracious at any time. The stomach can emulisfy and assimilate small quantities of food frequently taken. Whereas, if only three meals are consumed, and the appetite for breakfast does not permit the patient to eat food enough at that time to supply waste, the deficiency must be made up at subsequent meals. The result is an overloading of the stomach, and the over-repletion retards digestion, rendering it unnatural and labored. The amount of food assimilated from such a meal is entirely out of proportion to the quantity of aliment ingested.

The gastric glands are excited to undue activity, in an effort to furnish the requisite juices for the digestion of so large a quantity of food. At the next meal, the exhausted glands cannot supply the necessary secretion, and the vitally important gastric functions are not performed normally. Danger of future dyspepsia is thus incurred. The habit of drinking large quantities of liquid during a meal predisposes the system to dyspepsia. Before the act of digestion can be normally begun the aliment must be of proper consistency to be appropriated. A superabundance of liquid retards healthy action, and the excess has to be absorbed before digestion can proceed. Milk or water may be taken immediatly after a meal, with good effect. Also between the hours of active digestion, water can be liberally consumed with good results, as it facilitates the expulsion from the system of excrementitious waste, and corrects any tendency that may exist to accumulate solid effete matter in the blood.

The use of alcohol to urge on digestion should be positively interdicted. There is a paralysis of the stomach, that simulates dyspepsia, which attacks persons subsequent to their thirty-eighth or forty-fifth year of age, when the use of brandy as an aliment is indicated. The disease, arising from excesses and indiscretions in eating and hastily departing for some labor, usually physical, can only be cured by a systematic treatment involving an entire change of habits. A large proportion of persons in this condition will not appreciate the necessity for the employment of certain methods unless the treatment is surrounded by mystery. They will take medicine with wonderful regularity, but will neglect to follow important instructions regarding diet and hygiene. With patients who cannot control themselves, or remember instructions, it becomes necessary to prepare a bill of fare, and to note for their guidance full directions —step by step. If a chronic inflammation of the mucous coats of the stomach exists, the iodide potassium compounded with iodine will allay it in a large proportion of cases. When the condition is induced by an inherited tendency, tonic remedies, to build up the general system, may follow.

Pyrosis is often controlled by full doses of bismuth. Cardialgia responds to alkaline remedies. Such drugs should not be too freely used, however, when the patient is under constitutional treatment, as they are only palliative and not curative. It is not good practice to fill the stomach with drugs under any circumstances.

CHAPTER XXXII.

DYSPEPSIA COMPLICATED WITH MENTAL EXHAUSTION.—ITS TREATMENT.

WHEN the disease follows mental exhaustion and prostration, treatment must be directed not only to the correction of the physical disorder, but to the mind also. The last is difficult at best to approach. By sympathetic attention, however, the physician can often gain a patient's confidence, and in that case an important advantage is obtained. He should not ridicule the peculiarities developed by the disease, or make light of hypochondriacal fears, as the evils feared are real to the sufferer. The physician should endeavor to convince the patient of the unwarrantable nature of his fears. He should also soothe and tranquillize the patient by removing the cause of uneasiness. A lady applied to us a few months ago for advice respecting her husband's precarious condition, fearing that insanity had been induced by chronic dyspepsia, and liver complaint.

The man entertained a conviction that ruin and starvation were staring him in the face, when in reality he could retire from business with a competency. While giving him a preparatory treatment, we obtained from his son an exact statement of his business affairs, and then presented for our patient's consideration a hypothetical case, requesting him to favor us with his opinion regarding the situation as presented. He declared the condition and prospects of the suppositional person to be most favorable. On carefully informing him that we had simply shown him the state of his own business, he freely admitted that the facts gave no cause for anxiety. We assured him that his condition of mind was the legitimate result of disease, and that he must bear that fact in mind, and remain passive under treatment. He had also heard strange knockings on the floor above, and on the door. His family had innocently aggravated his condition, by humoring this fancy, saying that they would have the knocking stopped. By a direct and candid explanation, we succeeded in convincing him that the supposed knocking was simply a manifestation of his disease.

For days he would imagine that some one was trying to poison him. Again his family inadvertently kept alive and intensified these fears, by attempting to watch him surreptitiously, instead of explaining matters in full, and treating him as a rational being, compelling belief by truth and candor. By a thorough and frank explanation he was finally convinced that he was a sick man, and his family gladly changed their attitude towards him. His daughter entertained him by reading, and talking on pleasing topics—in lieu of suspiciously watching his movements. Daily rides in the open air were enjoyed. Under the combined influence of hygienic and medicinal measures, his mind slowly regained its equilibrium. The iodide potassium was administered for two weeks on beginning treatment, in the following prescription:

R̥.   Iod. Pot. ℥ iv.
        Hydrag-bin-iodide, ℥ iii.
        Ammonia iodide, ℥ ii.
                Aqua, ℥ vii.

M. Sig. One teaspoonful in wine-glass of water after meals. An enema of castor-oil was given at bed-time, and retained, until the bowels became regular, and his food was carefully chosen. Warm salt-water baths, followed by vigorous rubbing with coarse towels, were given immediately before he retired, in order to induce sleep. On withdrawing the first prescription, dilute phosphoric acid was administered in doses of ten drops, after meals.

A thin flannel bandage was also constantly worn over the abdomen. This was sprinkled night and morning with the following compound:

R̥.   Eucalyptus Tr. ℥ ii.
        Jaborandi Tr. ℥ i.
        Cinchona Tr. ℥ ii.
        Gentian com. Tr. ℥ ii.

M. A dry bandage was placed over the first to insure a uniform degree of warmth, and also to facilitate the absorption of the remedies which were intended by their tonic action to give tone to the liver and spleen, keeping their secretions relaxed.   In regulating his diet, we advised that he abstain from meat for a time, as he had always been a great meat eater, substituting milk in its stead.   His dyspeptic symptoms were allayed, and the. mental functions were restored to their usual condition. A trip through the west of two weeks' duration was then made, and he returned apparently fully restored.

Pepsin is highly recommended, but we have failed to obtain from its use the results described by writers, except when dietetic errors were corrected at the same time, rendering it a difficult matter to determine how much of the result is to be attributed to medicine and how much to regimen.   In paralysis of the stomach the action of pepsin is favorable, if it is given in large doses, say forty or sixty grains, in combination with milk and brandy.   It sustains the powers of life, until nature comes to the rescue.

Vomiting in dyspepsia is symptomatic, indicating morbid sensibility of the mucous coats lining the stomach. While waiting for constitutional effects—from remedies given to remove the cause—bisulphite of soda is suggested, and we have also used carbolic acid, one drop to the ounce of water.   The vomit usually contains a vegetable fungus, called sarcina ventriculi, and the acid destroys its activity.

Clothing should be warm, and especial effort should be made to maintain a uniform degree of heat over the abdomen changing the clothing as the weather requires, and not permitting the patient to make himself uncomfortable and to invite prostration through profuse perspiration, by wearing heavy clothing on a warm day, in anticipation of a sudden change.

Good health, and a normal condition of the gastric functions, demand that all the secretions and excretions be kept active. A sponge bath quickly taken each day, followed by vigorous rubbing with a coarse towel, until the surface of the body is red, goes far to secure that end. Individuals should discover the method of bathing best suited to their particular temperaments. A bath producing salutary results should excite a pleasing reaction. If the vital forces cannot sustain the shock of a cold water bath, warm or tepid water should be used, adding salt, mustard, or an alkaloid, as circumstances demand; but in all cases the first effects should be a sensation of renewed strength, with an invigorating glow of self-satisfaction. Each person should experiment until he finds what his system requires, as cleanliness is essential to good health, and is a prophylactic against disease, at all seasons of the year. There is a greater necessity for bathing, however, during the winter months than in the summer, as the secretions are kept active by continual transpiration in warm weather, and are not so subject to the danger of becoming torpid —inducing coryza, and colds—as they are in winter, if not stimulated to action by daily baths.

A certain amount of out-door exercise should be taken each day, to maintain a healthy condition, and to promote digestion. Gentlemen having inveterate shop-pers for wives, who go from shop to shop many times each week, may console themselves with the reflection, that their consorts seldom have dyspepsia. As dry goods' bills increase the liability to the disease decreases.

There is a condition simulating dyspepsia, which is often mistaken for it. It is excited by tænia, or tape-worm. No well-marked diagnostic symptoms are pres-ent. A patient presented himself for treatment in the spring of 1879. whose symptoms were indicative of dys-pepsia. He had been treated for that disease during the preceding ten months. We diagnosticated his case by exclusion as tape-worm, and relieved him of a worm forty-six feet long. The appetite in cases of tape-worm is capricious, being sometimes voracious, and often entirely wanting; colic is complained of at times, with a disagreeable sense of fulness and pain through the stomach. Tinnitus aurium, or roaring in the ears, with disturbances of vision, and vertigo, are suffered. Pruritus, or itching of the nose, is experienced. The desire for food in many instances cannot be controlled, and the patient consumes prodigious quantities, with no diminu-tion of appetite, at the same time becoming steadily more emaciated and weaker.

Persons who are suffering with this disease are unable to define their exact condition. It is a popular fancy, that food taken is devoured by the worm. In fact the worm derives its nutrition by imbibation. The loss of flesh by the patient, and the failure to receive benefit from the large amount of food consumed, are due to the loss of absorbing space through the presence of the worm. The food consumed cannot be appropriated for that reason, and the patient grows progressively thinner, while steadily increasing the amount of food consumed.

The treatment is simple and effective. We usually administer in the evening, salts, gamboge, or podophyllin to start the bowels. The following morning—before the patient takes food, we give one teaspoonful of Tr. male fern. It is well to omit breakfast, giving teaspoonful doses of fern every four hours during the day. At bed-time we administer:

R.   Castor-oil, ℥ i.
    Turpentine, ℨ i.
    Gum-arabic, ℨ iv.
    White-sugar, ℨ iii.
    Capsicum, gr. i.
  Calomel triturated, gr. ii.
M. Rub for twenty minutes in a mortar.

The worm will be discharged the following morning. If smaller doses than these are taken, the inconvenience of passing small sections of worm will be experienced. When turpentine or oil cannot be tolerated, or if the worm is not discharged by the first, we administer an emulsion of pumpkin-seeds, which will prove promptly efficient. The head of the worm should be searched for; yet in two cases we have failed to find it, and no return of the disease has occurred. The male fern acts by killing the tænia, and the remedies suggested expel it from the system.

We have noticed with patients suffering from dyspepsia and the chloral habit, a purulent ophthalmitis of a severe type, appearing as a sequel to the combined effects of disease and habit; and it will resist all measures of treatment, until the exciting cause is permanently removed, when it will readily yield to topical treatment, a weak solution of argenti nitræ, applied with a camel's hair brush, three times a day, dilating the pupils with the following:

℞   Atropine, gr. i.
                    Glycerine, ℨ iii.
                    Aqua rosc, ℥ ii.

M. Sig.   Put four to six drops of the above in either eye every other morning, one application being sufficient for the day.

The necessity for curing the chloral habit before attempting to treat complications of any kind is imperative.

CHAPTER XXXIII.

CHLOROFORM.—THE DANGERS INCIDENT TO ITS ADMINIS-
TRATION.—HOW THEY MAY BE MODIFIED.—ITS EFFECTS
UPON THE SYSTEM WHEN TAKEN HABITUALLY.—AND ITS'
TREATMENT AS A HABIT.

PRESCRIBING or giving consent to the inhalation of
chloroform, in periodical or other painful conditions, is attended with great danger A patient racked with pain is
soothed and relieved by its effects, and those who are familiar with its action are apt to indulge, under such circumstances, without the physician's consent or knowledge.
If fatal results do not follow, the patient, when forced
to discontinue its inhalation, will by his entreaties,
prayers and loud demands for its continuance, create a
scene never to be forgotten by those who have passed
through the trying ordeal.

A habit of using this agent can be avoided by prescribing the chloroform compounded in the following
manner. To every ounce of chloroform, add spirits
turpentine ℥ iv., nitrate of amyle ℥ ii. This modifies the
action of chloroform, without destroying its anæsthetic
properties. The turpentine and amyle excite a disagreeable nausea, if the inhalation is continued beyond a certain length of time, causing the patient to desire its discontinuance. It may be well to mention in this connec-

tion a discovery lately made, demonstrating that turpentine inhaled with chloroform in surgical operations protects the patient from those suddenly developed symptoms of collapse, that will unavoidably occur at times during the inhalation of pure chloroform. The turpentine acts through the vaso-motor system of nerves, giving strength and tone to the muscles presiding over respiration. We regret that we have not the reference to give.

This powerful anæsthetic is resorted to by a numerous class in the hope of allaying pain, or drowning real or fancied trouble. As a stimulant to satisfy the cravings of a depraved appetite, it is seldom taken. Yet, such cases sometimes present themselves, a large proportion of the victims being women. When a habit of using chloroform is cultivated, it seems to create an uncontrollable desire for its repetition at certain times. Tissue changes are not of such a character as to maintain or to constitute a physical demand. Decided changes, however, take place when the practice is continued, there being an arrest of molecular activity. The red corpuscles do not carry the amount of oxygen requisite to a normal degree of functional activity, or return to the lungs, for exhalation, carbonic acid gas, as it accumulates.

The victims become pale and asthenic, and general motor and sensory neuræmia is superinduced. The encephalon does not receive the blood necessary for a normal degree of physical and mental activity. They lose the power of mental concentration, and the ability to exercise their reason. They do not appear to realize the danger incurred, when they saturate a handkerchief with chloroform, on going to bed, and hold it over their nostrils until they are stupefied. Victims to the drug are frequently discovered dead in their beds, with a cloth clutched in the hands, and partially held in the mouth and hands. Death results from pulmonary asphyxia, preceded by muscular contractions. After death the brain is found congested, the mucous membrane of the air passages being in the same condition. Blood tinged with mucous is generally present in the bronchia and stomach. The autopsies are usually made some hours after death, and we cannot say how many of the changes are post-mortem.

Many victims have a habit of inhaling chloroform for several consecutive days, only stopping when warned by peculiar symptoms that death is approaching. They indulge periodically in a chloroform debauch, and a fearful spectacle is presented after they have inhaled the chloroform for several hours, consuming just enough to maintain a drunken state. The face has a corpse-like appearance, the lips being bloodless and blue, and covered with a frothy saliva, tinged with blood, making the victim look ghastly in the extreme. The eyes are congested and expressionless, with dilated pupils. Respiration is slow and stertorous. Blood is often passed at stool, and frequently vomited. Articulation is difficult and forced. The victims are often seized with muscular spasms—becoming rigid and cold, remaining in a cataleptic state for many minutes. On being restored to consciousness, if chloroform is not administered at once, or is refused, they seem to forget their surroundings, and all sense of shame; leap out of bed whether clothed or not, and beg for more in a hysterical manner—on their knees, if necessary, resorting to any expedient to gain their end. Many ladies have become familiar with the pleasing effects of chloroform in the lying-in room. (We will hereafter suggest the preparatory treatment to ameliorate their pains, enabling the physician to dispense with the use of anæsthetics.) Others inhale its vapor to induce sleep; an increased quantity soon becomes necessary, and the desire to resort to its use during the day gradually engulfs them. Desperate risks are taken, as they usually seek a secluded spot in which to indulge their habit, greatly enhancing the danger of syncope, by being alone and in an upright position. Persons fainting from any cause, should be placed with the head in a dependent position, to facilitate the return of arterialized blood to the brain: otherwise the brain rapidly becomes anæmic, producing paralysis of the muscles which preside over respiration.

As persons addicted to the use of chloroform are passing into a state of collapse from its inhalation, they give utterance—involuntarily—to a prolonged groan, peculiar in tone, similar in quality and intensity to the moans of persons suffering from night-mare. These sounds alarm friends, giving them an opportunity to save many lives, that would otherwise be sacrificed.

When chloroform is inhaled to induce sleep, its immediate discontinuance is indicated. Natural rest may be induced by the means suggested for use with opium eaters. If the habit is indulged during the day, it is well to mix nit-amyle and turpentine with the chloroform on hand; if that is done the patient will voluntarily discontinue its use to avoid nauseating results. When the patient is threatened with asphyxia, from the use of poisonous quantities, amyle turpentine, or carb-ammonia, by the stomach, enema, or subcutaneously, is demanded. The head must be kept low and surrounded by warm cloths, while the face is sprinkled with cold water, to excite reflex action, by transmitting tone through the cutaneous nerves, to the central functions. The fumes from ammonia act favorably, and keeping a current of air passing over the face, by fanning or otherwise, is of advantage. Extreme measures consist in giving the plunge bath, creating artificial respiration, and placing the extremities in cold water, while the head is treated to warm water packs. Cupping the spine and blistering the back of the neck materially assist in establishing a normal condition. The juice of lemons, brandy and spirits ammonia, with warm soups, may be given as soon as possible. The above described treatment is also applicable where over-doses of chloral have been taken.

CHAPTER XXXIV.

THE INCREASED PAIN OF THE PUERPERAL STATE, DEMAND-
ING ANÆSTHETICS.—HOW TO AVOID THE USE OF ANÆS-
THETICS IN THE LYING-IN ROOM, BY RENDERING THE
PAINS OF LABOR MORE BEARABLE.

By controlling the intensity of pain, and rendering
bearable the dreaded agony of the lying-in room, we
not only strike at the root of the necessity for administering anæsthetics, but also remove a temptation to commit a terrible crime—a sin that has become so common
as to be regarded lightly by large numbers of persons—
the crime of abortion. How to overcome the pain of
parturition, is a question that contributors to current
medical literature have strangely neglected to discuss
in that aggressive spirit which is characteristic of the
American physician.

The important matter to consider, and that to which
all others are subsidiary, is the control of pain. The
word "pain," as we usually understand its significance,
suggests but a faint notion of that fearful agony which
is sustained by our American mothers. It is a grinding,
tearing pang, intensified by the knowledge that it must
be borne, and that there are no avenues of escape, no re-
spites, whether the parts involved are physically adapted

to the inevitable trial or not. The babe must force its way through its pain-riven and blood-strewn path. To escape from this agony, women of civilized nations resort to chloroform, with its immediate dangers and the certainty of misery which its use entails if a habit is formed.

This intense pain in parturition drives young mothers diligently to seek for every means of escape from a second experience, inciting them to commit, or give their countenance to a dastardly murder—the wilful slaughter of a defenceless being, whose right to enjoy unmolested its uterine life is as certain and sacred as is its right to life when, a few months later, it assumes an independent existence.

Outraged nature holds women who commit this offence to strict account, the unnatural crime being productive of serious injury to the constitution, blasting in many cases the bud of health before it can bloom into mature womanhood, and often planting the seeds of fatal disease, if it does not cause immediate death. As for the man who slays defenceless innocents, the title physician should not be prostituted by coupling it with his name.

The known abortionist excites the horror and just contempt of his fellow men. Admiration may sometimes be mingled with pity for the daring highwayman, who commits murder in carrying out his projects of gain by violence. He knowingly meets those who are prepared for self-defence, by age and discretion.

The embryonic life makes no resistance, and the hands intent upon murder accomplish their work of blood without risk, and a life goes out without a groan.

The conscientious physician should labor unceasingly to modify this terrible and unnatural pain, and so remove the first temptation to commit the crime, by placing the civilized mother's system in a condition similar to that of her wild and untutored sister, who entertains no dread of parturition, which to the savage woman is comparatively a painless act of nature.

It is within the power of the profession to mitigate, if not entirely to relieve women from the throes of labor. That fact has been satisfactorily determined by actual demonstration; and the only barrier standing in the way of the accomplishment of this result can be removed by the mother, if she will not wait for the eleventh hour before calling the doctor, but by timely warning will give him an opportunity to carry out a preparatory treatment, and prepare her for the trying ordeal.

South Sea Island women, who are by force of circumstances confined to a fruit diet, experience few premonitory symptoms, and labor is quickly and painlessly accomplished. Going for a walk, or to visit friends in the morning, if overtaken by symptoms of labor, they stop a few minutes, give birth to a robust babe, and return home with the infant in their arms. English ladies, judiciously treated to prepare them for a painless labor, have given birth to well-developed babes, without consciousness of the fact, and while enjoying natural sleep. These assertions may appear extravagant, but they are well authenticated, and many more illustrations could be cited, if that were necessary, to establish the fact.

The measures to be adopted include a selected diet, and hygienic observances. The pregnant woman should not be subjected to the drudgery, and consequent anxiety of household duties. She should take exercise that will call every muscle into play, inducing a uniform degree of development, and so prepare the body to endure, with the least possible pain and injury, the unusual exertion to be thrown upon it. Exercise may be taken each day, by walking, the use of light wooden dumb-bells and rowing in summer, or using the health-lift in winter.

Corsets, or other closely fitting articles of dress, should not be worn after the first month of gestation, and excesses of all kinds must be avoided. The surroundings must be pleasant, and pregnant women should be free from mental worry, and have no occasion to get out of humor. Amusements and agreeable company are essential.

Bathing regularly in salt-water, and especially using the sitz bath, is highly important, the temperature of the bath being high or low, whichever excites an agreeable reaction. Under no circumstances are they to indulge in alcoholic drinks or narcotics, for their stimulating effects.

Their fare may include all that is included under the

name of a fruit diet. They should particularly avoid all food containing a large percentage of earthy, or calcareous matter, as a large proportion of such food goes to supply the brittle, unyielding element in bone.

A non-calcareous diet renders the frame work of the fœtus yielding and pliable, enabling it to glide through the pelvis easily, by conforming itself to the parts. Wheaten flour should be avoided, as calcareous matter is largely an ingredient. Spring water also is heavily charged with lime, and instead of it pregnant women should drink, if possible, filtered rain-water, cooled with ice.

The diet may be abundant and nutritious, but it should be confined to fruit, vegetables, game, and occasionally fish, embracing also the juice of a lemon before breakfast, or if lemons are out of season, two oranges can be substituted. Rice, sago, raisins and figs may be freely taken. Articles of food necessary to a perfect organization of the babe's osseous structure and teeth can be liberally consumed by the mother after labor, and will be rapidly appropriated by the child. The diet suggested, with its beneficial tendencies, secures the enjoyment of perfect health to the mother, and the fœtus draws freely upon her surplus vitality—presenting an excellent foundation to build upon.

By such a regimen a painless parturition is not only approximated, but dreaded after-complications that strike terror to the heart of physicians and friends—such as puerperal fever, and phlegmasia dolens—are en

tirely avoided, as the diet has facilitated the free elimination of those humors that excite inflammatory action,
during the puerperal state.

We have observed the gratifying results of this treatment, with a lady who, owing to peculiarities of temperament, was sadly in need of its salutary effects, as the
period of gestation with her was one of great suffering
and general ill-health, beginning with obstinate constipation, which was followed by gastric irritability, the
woman becoming sallow, emaciated and anæmic. The
lower extremities became extremely swollen and painful,
the veins full and prominent, and the patient found no
relief until after she had suffered severely with a prolonged labor.

With her last two children, this patient willingly con-
formed to the treatment suggested, taking the juice of one
lemon on rising, and eating two oranges during the day,
with fruits and vegetables as directed. Her bowels became regular, and the functions of the skin remained
active, keeping her complexion clear. The painfully
swollen condition of the lower extremities did not appear her general health being greatly improved. The
nine months were passed without a pain or complication
of any description, and the woman gave birth to a finely
proportioned babe within two hours after experiencing
faint premonitory symptoms. She afterward declared
that she had suffered more acutely in former pregnancies,
from constipation and its results, than she had suffered
this time in labor.

In formulating a plan of treatment, the physician should keep in view the requirements of each patient, as suggested by temperamental peculiarities, if he would secure entire success.

CHAPTER XXXV.

BROMIDE OF POTASSIUM.—ITS DEMENTING EFFECTS WHEN TAKEN HABITUALLY.—ITS TREATMENT AS A HABIT.

DURING the last fifteen years it has been the custom with many physicians, to prolong the use of bromide of potassium with patients suffering epilepsy and insomnia. Its continual use produces a train of evils, of a decided character—disastrous to the mental capabilities ; and to prolong the administration of this drug from year to year is to consign the patient to a condition approaching idiocy. After years of assiduous dosing with the drug, in the hope of curing epilepsy, countless victims are left by its use weak in brain-power, with shattered constitutions, and predisposed to acquire vile and degrading habits. And what are the results? Eminent authorities, such as Stille and Maisch, after weighing the combined results obtained by our best observers, say : "Whether or not radical and absolute cures are ever made of genuine epilepsy is undetermined by the use of bromide potassium."

After carefully surveying the variable results obtained we consider the drug curative in a small proportion of cases, and favor a judicious trial of its merits in each individual case of epilepsy. We have noticed when favorable results follow its administration, that the disease usually responds to its action during the first two months.   If no impression is made within that time, and the convulsions are as frequent and severe as at first, its further use is not indicated and will only cause complications.

The physician can discontinue its use after such a trial, without imperilling the patient's after-health.   Its physiological action is more satisfactory when combined with bromide of ammonia.   Equal parts may be given during the trial.   The combined effects are less depressing to the heart's action.   The anti-spasmodic properties of the mixtures are equal, if not superior to those of the potassium, and it attains its maximum of effect much more quickly.   We give the ammonium the preference in general practice.

The anaiphrodisiac properties of bromide of potassium are greatly over-estimated, as long experience with the drug demonstrates.   While the system is laboring under its first direct action, as a depresser of the capillary circulation, it may control sexual desire for a limited time, but it is a fact familiar to those who have epileptic asylums under their control, that masturbation is practised to an uncontrollable extent by the inmates, taxing the ingenuity of physicians in charge to devise a remedy; and yet the patients are regularly taking enormous doses of the drug.

Under its habitual use, the breath has a peculiar and offensive odor.   The cuticle becomes thick and discolored, with pustular papules, that disfigure the face. Striking effects are observed in the action of the drug upon the brain and spinal cord; it interferes with the mutrition of these organs, causing a progressive loss of memory. The mucous coats are red and congested, owing to a contracted condition of their capillaries. The muscles become enfeebled, and constipation, with a scant urinary secretion, is usually present.

The appearance of diarrhœa is generally followed by death, as the remaining strength of the patient is soon exhausted. Tactile sensibility decreases, and sight and hearing lose their acuteness. Exceptional cases present hallucinations, leaving the patient a wreck in mind and body, and forcing the conclusion upon us that the remedy is more to be dreaded than the disease.

Such facts should not, however, deter the young physician from judicious use of this drug, but should excite caution. Unlike opium, it can be stopped at once without creating decided nervous troubles, insomnia being the principal symptom which demands its continuance. On its withdrawal, persons who have used the drug experience an indescribable feeling of heaviness about the head, with a confusion of ideas, completely incapacitating them for well-directed mental effort. Epileptics will display symptoms of insanity for a few days after the withdrawal of the drug. Their convulsions are also increased in severity and frequency, followed by symptoms of improvement. The latter should not encourage friends, however, as such persons do not maintain their improved condition for any considerable length of time after reaction has taken place. The complexion will clear, and the breath will lose its offensive odor. In fact a general change for the better is observable.

Patients who have formed the habit while endeavoring to allay sleeplessness should be cautiously treated. The physician should discover the exciting cause that originally induced such a state, and remove it by appropriate measures, correcting the patient's dietary and hygienic indiscretions at the same time. A constitutional treatment, with tonics and cautery, will be indicated if he is anæmic, or has sustained prolonged mental strain. Otherwise to withdraw the drug is to invite obstinate insomnia, that reflects its debilitating effects throughout the economy.

After putting the system in the best possible condition, stop the drug at once, and induce sleep, in the manner suggested for use in the case of victims of opium. Musk and cannabis indica may be given in large doses, and depended upon to allay undue irritability.

We treated a brilliant young minister, who first used the drug to control a desire for liquor. He had gone to great excesses, taking two and three drachms per day. The bromide was withdrawn and the following preparation administered:

℞.   Cinchona rubra, ℥ iv.
                           Cannabis indica, ℥ ii.
                       Gentian comp. Tr. ℥ ii.
                              Ginger, ℥ i.
                       Capsicum, ℨ ii.

M. Sig. Two teaspoonsful three times a day. Musk and phosphoric acid were given at bed-time, after a warm bath had been taken. We have given chloride of gold occasionally, with very good results, not only in his condition, but in cases of irritability following acute alcoholism. It cannot, however, be depended upon in either case.

Our patient resigned from the ministry a few months before coming under observation, because of his uncontrollable appetite for strong drink, which haunted him continually. Being a terse and ready writer, he assumed full charge of the editorial department of a thriving journal, accomplishing the arduous task imposed with ease and acquitting himself with great credit. The drug by degrees incapacitated him to such an extent, that he found it utterly beyond his power to produce simple editorial articles, his intellectual faculties being blunted. He was finally obliged to abandon his position. After a prolonged treatment, embracing the measures referred to above, and including the benefits derived from a sea-voyage, his mental capabilities resumed their normal condition.

Complications pre-existing, or occurring as concomitants of this state, may be treated as if presented under other circumstances.

Great mistakes and complete failures are often made, by attempting to cure a habit, without removing the cause which first prompted its formation. We refer to this important matter in treating the dementing habits to indelibly impress upon the mind by frequent repetition the necessity for care.   As serious consequences will inevitably follow neglect of this caution, ending finally in the patient giving up in despair; and falling back into the opium, liquor, chloral or bromide potassium habit.

CHAPTER XXXV.

CANNABIS INDICA.—ITS WONDERFUL TONIC PROPERTIES.—
MEDICINAL EFFECTS UPON THE SYSTEM.—ITS HABITUAL
USE AND RESULTS.

OUR pharmacologists do not give this drug the credit
and prominence as an anti-spasmodic and general tonic
which our experience leads us to believe that it deserves.
We find that in diseases which are characterized by
failure upon the part of the assimilative processes to
provide the economy with a sufficient amount of nutrition, and where it is important to reawaken and sustain
those functions in their highest efficiency, the favorable
effects of cannabis indica are incontrovertible.

When the general system from any cause is weakened
and relaxed, having sustained the loss of tissue, and the
peripheral vessels, for the want of nerve power, have
lost their tone, the cannabis indica will prove an
efficient agency in imparting tonicity, enabling the system to improve its nutrition. That the drug exerts a
powerful influence over the endangium, is exemplified
in atonic conditions of the uterus; in such cases it
transmits tone and force to the uterine functions—facilitating labor and controlling hemorrhage. It should not
be administered as a soporific, depending upon its direct

action. When general poverty of the system exists, exciting insomnia through the irritability that is apt to accompany progressive anæmia, we derive a soporific action from it, in a reflex manner.

In this country we cannot obtain from the drug the effects described by writers in foreign countries. It has proven itself to be highly susceptible to changes incident to climate, transportation and handling. We therefore cannot accept accounts of its action in its native clime, India or China, as indicative of the effects it will work if used here.

Travellers favor us with vivid and highly interesting descriptions of their experiences with *hasheesh,* or *Indian hemp.* In Arabia, India and China, its anæsthetic effects are so decided as to admit of surgical operations being performed while the subject is under its influence.

Bayard Taylor, Humboldt, and other conscientious observers, describe a peculiar effect produced by large medical doses—namely, the multiplying many times of the size of surrounding objects. A loaf of bread appears to a person so drugged like a small-sized mountain. The drug exercises no such influence in this country over the optic nerves; neither will it induce anæsthesia. A young druggist came to our office greatly frightened, over peculiar symptoms excited by fifteen grains of solid extract of cannabis indica (the usual dose is one-third, one-fourth, or one-fifth of a grain). His face was flushed and swollen, being dry and feverish. The extremities were hyperæmic, with pulse regular, full and bounding; respiration was quickened but free. The eyes were congested and the pupils dilated. The temperature was from one to two and one-fourth degrees above the normal heat of the body. In protruding the tongue, spasmodic effort kept it in motion.

He described his feeling as anything but pleasant. A sinking, fainting sensation, seeming to arise from the stomach, threatened syncope, and then passed off suddenly. A marked symptom was a feeling as if an iron band tightly surrounded the head. We afterwards observed the latter symptom in the case of one woman and two men, who had taken overdoses of cannabis indica. The general appearance was similar with all these cases.

The intellectual faculties are stimulated by this drug to a certain degree, but not so actively as by opium. The effects of a large dose of cannabis indica resemble those obtained from nitrate of amyle, especially in the action of the drug upon the capillaries of the face. Its disagreeable effects when taken in excessive quantities, counterbalance any pleasing sensations, a fact which lessens the danger of forming a habit of its use, and leaves no incentive to its abuse. There are unfortunately exceptions to this, where the action of the drug fascinates the devotee, influencing him to commit fearful excesses ending in extreme debility. The mind becomes imbecile, and ultimately death by marasmus ensues. In order to obtain the fine medicinal effects of cannabis indica as a nerve and general tonic, doses of one-

eighth to one-fourth of a grain, according to the susceptibility of the patient, may be administered, conferring
upon the system its beneficial effects, without exciting
its stimulating action.

If overdoses are inadvertently taken, the juice of
lemons will counteract alarming symptoms. We have
prescribed the cannabis indica in combination with
other tonics, in kidney complications, including albu-
minuria and nephritis. It supports the economy by
facilitating the assimilative processes, giving tone to
the glands of the skin which secrete water, urea and
salts, and upon which so much depends, in the last
named disease, in neutralizing the carbonate of ammonia in the blood. During labor it is equally efficacious in promoting uterine contractions when their
cessation imperils both mother and child.

In diseases of the lungs, where tissue waste is rapid,
its tonic influence is strikingly apparent. The same
holds true in prostrating diseases that undermine the
system, leaving the patient weak and anæmic. The
system in low typhoid conditions responds to its action,
and uniformly good results follow.

Great difficulty in obtaining a pure extract of cannabis indica is often experienced by physicians. It is of
the utmost importance that the best quality should be
administered, as a low grade is utterly worthless as a
medicine. We are persuaded that the difficulty is due
to adulterations and faulty and damaged crops being
used, and this perhaps explains the fact that so many

physicians declare that only negative results have been obtained by them. Our best manufacturing chemists say that the crop is uncertain and easily damaged. We use an English solid extract, and our druggist makes an alcoholic tincture by dissolving one grain of the extract in each drachm of alcohol. In writing for it alone, we order it compounded in gum-arabic or glycerine.

If the cannabis indica is of standard quality and in proper condition, it will not, when compounded as above, precipitate, or break from a perfect solution, when re-compounded with water, glycerine, fluid extracts, or tinctures. If it should precipitate it is best not to prescribe it, as doubtful results will follow. If the question of suitable brands did not involve the well-being of those who depend upon us for relief, we would not have referred to any special brand. And in so doing we do not question the purity of any other.

www.ingramcontent.com/pod-product-compliance
Lightning Source LLC
Chambersburg PA
CBHW021517210326
41599CB00012B/1289